AF356459

# Biomechanics Energetics of Natural Assisted Human Comparative Movement Locomotion

# Biomechanics Energetics of Natural Assisted Human Comparative Movement Locomotion

Editor

**Luca Paolo Ardigo**

MDPI • Basel • Beijing • Wuhan • Barcelona • Belgrade • Manchester • Tokyo • Cluj • Tianjin

*Editor*
Luca Paolo Ardigo
University of Verona
Italy

*Editorial Office*
MDPI
St. Alban-Anlage 66
4052 Basel, Switzerland

This is a reprint of articles from the Special Issue published online in the open access journal *Symmetry* (ISSN 2073-8994) (available at: https://www.mdpi.com/journal/symmetry/special_issues/Biomechanics_Human_Comparative).

For citation purposes, cite each article independently as indicated on the article page online and as indicated below:

LastName, A.A.; LastName, B.B.; LastName, C.C. Article Title. *Journal Name* **Year**, *Volume Number*, Page Range.

**ISBN 978-3-0365-0414-8 (Hbk)**
**ISBN 978-3-0365-0415-5 (PDF)**

Cover image courtesy of Luca Paolo Ardigo

# Contents

# About the Editor

**Luca Paolo Ardigo**, Assistant Professor in Sport Science, is a physiologist and biomechanist at the School of Exercise and Sport Science, Department of Neurosciences, Biomedicine and Movement Sciences of the University of Verona in Italy. His main areas of research/teaching are the biomechanics and bioenergetics of natural and assisted locomotion and portable devices for measuring metabolic expenditure and physical activity. He first dealt with biomechanics and bioenergetics of assisted movement and locomotion during his Ph.D. in Biomechanics at the Exercise and Sport Sciences Department, Institute for Biophysical and Clinical Research into Human Movement of Manchester Metropolitan University in the UK (thesis title: "Energetics and mechanics of passively assisted locomotion"). He is the author of more than one hundred articles in scientific journals (32 of which as well as 1 book are on biomechanics and bioenergetics of human assisted movement and locomotion) and edited some of them. Additionally, he leads an international group for the design, construction and evaluation of "handbike" boats for disabled people.

# Preface to "Biomechanics Energetics of Natural Assisted Human Comparative Movement Locomotion"

Movement and locomotion have always been key activities for all animals, being related to the most crucial life functions: retrieving food, facing environmental issues and mating. Thus, humans are able to execute fully developed complex upper arms movements and bipedal gaits in order to move and locomote. To further enhance their performance, they started inventing smart passive mechanical tools. This need arose from both intrinsic limitations of their muscle–joint–bone systems and metabolic power availability. Newly invented devices were mainly introduced in order to cope with such constraints. Some of the potential topics and questions regarding symmetry and biomechanics and energetics of passively assisted human movement and locomotion include: how symmetrical/asymmetrical are most performing/economical human-powered assisted locomotion modes on land, water, and air; how did passively assisted locomotion and/or movement (for catching, manipulating, carrying, etc.) evolve over history in terms of symmetry degree; and how symmetrically do man-passive tool complexes best adapt to extreme environmental conditions (viz. different gradients, surfaces, media, etc.). The aim of this Special Issue is to advance knowledge regarding symmetry and biomechanics and energetics of passively assisted human movement and locomotion.

**Luca Paolo Ardigo**
*Editor*

*Article*

# Effects of Nordic Walking on Gait Symmetry in Mild Parkinson's Disease

Ana Paula J. Zanardi [1], Flávia G. Martinez [1], Edson S. da Silva [1], Marcela Z. Casal [1], Valéria F. Martins [1], Elren Passos-Monteiro [1,2], Aline N. Haas [1] and Leonardo A. Peyré-Tartaruga [1,*]

[1] Exercise Research Laboratory, School of Physical Education, Physical Therapy and Dance, Universidade Federal do Rio Grande do Sul, Porto Alegre 90000-000, Brazil; anapjzanardi@gmail.com (A.P.J.Z.); flavia.martinez@ufrgs.br (F.G.M.); edsonsoaressilva@hotmail.com (E.S.d.S.); marcelazcasal@gmail.com (M.Z.C.); valeria.feijo@ufrgs.br (V.F.M.); elren.monteiro@gmail.com (E.P.-M.); jhham5969@naver.com (A.N.H.)

[2] Laboratory of Locomotion PhysioMechanics, Faculty of Physical Education, Universidade Federal do Pará, Castanhal 68740-000, Brazil

* Correspondence: leonardo.tartaruga@ufrgs.br; Tel.: +55-51984063793

Received: 30 September 2019; Accepted: 2 December 2019; Published: 4 December 2019

**Abstract:** Individuals with Parkinson's disease (PD) have gait asymmetries, and exercise therapy may reduce the differences between more and less affected limbs. The Nordic walking (NW) training may contribute to reducing the asymmetry in upper and lower limb movements in people with PD. We compared the effects of 11 weeks of NW aerobic training on asymmetrical variables of gait in subjects with mild PD. Fourteen subjects with idiopathic PD, age: 66.8 ± 9.6 years, and Hoehn and Yard stage of 1.5 points were enrolled. The kinematic analysis was performed pre and post-intervention. Data were collected at two randomized walking speeds ($0.28~\mathrm{m\cdot s^{-1}}$ and $0.83~\mathrm{m\cdot s^{-1}}$) during five minutes on the treadmill without poles. The more affected and less affected body side symmetries (threshold at 5% between sides) of angular kinematics and spatiotemporal gait parameters were calculated. We used Generalized Estimating Equations with Bonferroni post hoc ($\alpha = 0.05$). Maximal flexion of the knee ($p = 0.007$) and maximal abduction of the hip ($p = 0.041$) were asymmetrical pre and became symmetrical post NW intervention. The differences occurred in the knee was less affected and the hip was more affected. We concluded that 11 weeks of NW training promoted similarities in gait parameters and improved knee and hip angular parameters for PD subjects.

**Keywords:** locomotion; biomechanics; pole walking; more affected side; asymmetries; angles; spatiotemporal; aerobic training

---

## 1. Introduction

Individuals with Parkinson's disease (PD) walk slower than able-bodied people. The lower segmental and trunk excursion are associated with a reduced stride length and increased double support time and stride variability during walking in PD subjects [1–4]. Additionally, the contralateral gait asymmetry is associated with the chance of a person with PD developing freezing of gait [2,5].

The literature shows that in people with PD, the tremor reduced the upper and lower limbs asymmetry, mainly at higher walking speeds [1]. During the gait cycle, the angular kinematic parameters, such as shoulder and elbow movement, hip flexion and extension, pelvic rotation, knee flexion, plantar, and dorsiflexion of ankle need to be coordinated to conserve energy. The preservation of the range of joint motion results in lower vibrations and lower impact forces during the gait as well as minor compensations during the task. Therefore, the symmetry during gait may result in lower energy expenditure [6–8].

Aerobic exercises improve the functionality, mechanics, and energetic parameters in people with PD [9–11]. In this context, Nordic Walking (NW) is an exercise modality that revealed functional improvements in older people with PD [12–15]. The NW is characterized by the use of poles that requires symmetrical and coordinated movements provided by arm participation to move the body forward [16]. Also, the range of upper limb motion is increased using poles [17] changing the muscular synergies, particularly the spatial organization [18] and the magnitude of activation [19] of upper limb muscles in comparison to free walking (FW) [18].

Although the natural history of illness suggests an increasing contralateral asymmetry in PD [4,20,21], the findings are controversial. For example, Delval and colleagues did not observe gait asymmetries [22], whereas the literature showed that the swing time are markedly differently between feet [4]. The asymmetries attributed to cardinal symptoms of PD seem to denote a natural functional difference between the limbs, particularly associated with propulsion and control tasks [23].

While the NW improves functional mobility and independence for PD [13,14], the potential improvement on the contralateral asymmetry after NW intervention remains unknown. Therefore, this study aimed to characterize the contralateral gait asymmetries and to investigate the effects of NW on these asymmetries in individuals with PD. We hypothesized that the kinematic variables during gait at pre-test period should be different between the sides (asymmetric), and that at post-test period, these differences should decrease, resulting in a more symmetrical gait.

## 2. Materials and Methods

### 2.1. Experimental Design

This is an experimental study that evaluated an intentional and non-probabilistic sample of people with Parkinson's disease. We first characterized the gait asymmetry and after tested the NW in the asymmetrical variables. We conducted the study in line with the protocol approved by the Ethics Committee of Research involving human beings from Universidade Federal do Rio Grande do Sul (project number: 69919017.3.0000.5347 and clinical trials ID: NCT03860649). All subjects gave their informed consent for inclusion before they participated in the study.

### 2.2. Participants

We included people with the diagnosis of idiopathic PD, aged 50 to 80 years, 1 to 3 on the Hoehn & Yahr (H&Y) scale [24], and physically inactive for at least one month. They should be in medical treatment and should have the ability to understand the verbal instructions to performing the tests. The exclusion criteria are: they should not have history of vertigo, surgeries in lower limbs during the last year, and making use of prostheses in the lower limbs. Further, the participants who undergone deep brain stimulation surgery, who had severe heart diseases or other associated neurological diseases, dementia, and who did not have conditions of ambulation and who got Montreal Cognitive Assessment (MoCA) less than 21 points were excluded [25]. Subjects who missed more than 75% of classes were excluded. Also, only ambulatory individuals and individuals who were walking without aid on the treadmill were included.

Calculation of the sample size was carried out using the Gpower v.3.1 program. Values of maximal flexion of the knee and maximal abduction of the hip from the study of Ribeiro et al., (2018) were used for the calculation [26], with a level of 5% and a power of 85%. A number of 14 subjects was estimated, considering the possible sample losses and a good adhesion rate estimated at 70% [27].

### 2.3. Assessment and Intervention Procedures

All procedures were carried out at the in the Biodynamic sector of the Exercise Research Laboratory (LAPEX). Subjects attended three distinct moments to perform data collection. On the first day, previous evaluation of the individual was performed to verify whether it fit the eligibility criteria. After these initial procedures, the Unified Parkinson's Disease Rating Scale (UPDRS) was performed to determine the side more and less affected [28] and H&Y scale. Subsequently, after 10 min of rest, the individual performed the six-minute walking test (6MWT) on the ground. After 10 min of rest, they were familiarized to the treadmill (INBRAMED, model ATL-Inbrasport, Porto Alegre, Brazil), and rating of perceived exertion (Borg Scale) for 15 min [15].

The kinematic analysis was performed pre and post-intervention of NW. The subjects walked in randomized walking speeds of 0.28 m·s$^{-1}$ and 0.83 m·s$^{-1}$ for three minutes, and the kinematic data collection was performed at the last minute. The kinematic data collection was carried out by the three-dimensional motion analysis system Vicon (Vicon Motion Capture System-Oxford Instrument Group-USA, 1984), using six infrared cameras (100 Hz, three cams Bonita with a resolution of 1 MP, and three cams T10 with a resolution of 1.3 MP). Thirty-five reflective spherical markers were placed on anatomic landmarks of interest according to the model Plug-in-Gait Full-Body. The three-dimensional reconstruction of the captured kinematic data was obtained automatically by the Vicon NEXUS®1.8.5 software. The system captured a filming space of 4 m wide, x 6 m long, and 3.5 m high.

The training period lasted 11 weeks, with two weekly sessions, being four sessions of NW technique adaptation and 18 sessions of NW training (22 sessions in total). The volume was determined by the session time in minutes and by the percentage of the distance covered in the 6MWT, which was determined individually for each subject (Equations (1)–(3)) [29]. Therefore, the individuals were divided in three groups of volume coefficient based on the individual capacity of the subjects. These groups include A1: those who walked at 50% of the 6MWT (coefficient less than 0.85), A2: those who walked at 70% of the 6MWT (coefficient between 0.86 and 1.2), and A3: those who walked at 100% of the 6MWT (coefficient above 1.2) in the first session. The intensity was based on the subjective intensity of gait that was comfortable, intermediate, maximal, and jog. Comfortable velocity is that speed that person normally walks in the street. The intermediate velocity is the speed between the comfortable and the maximum, and the maximum speed determines how fast as possible the person can walk without running, while the jog is the intensity in which the individuals will run for a short period.

$$\text{Volume coefficient = performed distance / predicted distance} \tag{1}$$

$$\text{Man: predicted distance (m)} = 493 + (2.2 \times \text{height}) - (0.93 \times \text{weight}) - (5.3 \times \text{age}) + 17\ \text{m} \tag{2}$$

$$\text{Woman: predicted distance (m)} = 493 + (2.2 \times \text{height}) - (0.93 \times \text{weight}) - (5.3 \times \text{age})\ \text{m} \tag{3}$$

The training periodization (Figure 1) was based on [15] and more details can be observed in supplementary material 1 (Tables S1–S3). We conducted the NW training on the athletics track (400 m) of the School of Physical Education, Physical Therapy and Dance (ESEFID) of the Universidade Federal do Rio Grande do Sul (UFRGS).

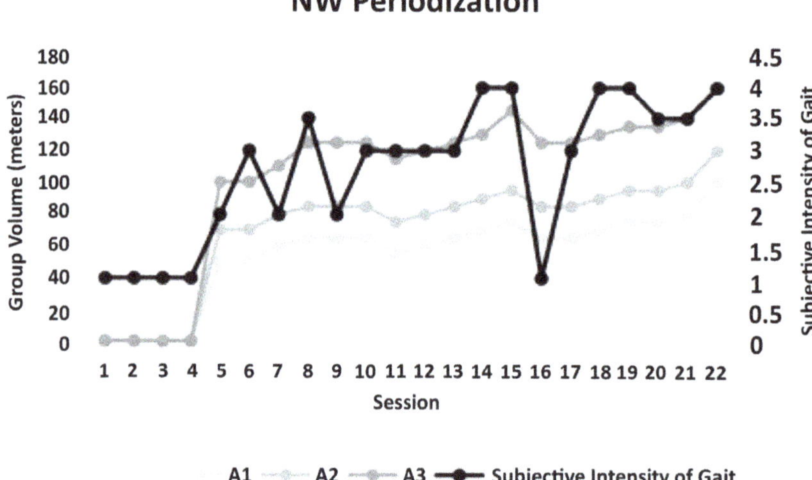



**Figure 1.** The training volume (grey lines, left vertical scale) is in meters, and intensity (black line, right vertical scale) is in rates of perceived speed during 22 training sessions. The subjective Intensity of Gait is the rate of perceived speed, and the number 1 represents (comfortable), 2 (comfortable and intermediate), 2.5 (intermediate and fast), 3 (comfortable, intermediate, and fast), 3.5 (intermediate and fast) and 4 (comfortable, intermediate, fast, and jog). The individuals included in the Group A1 are those who walk at 50% of the 6MWT (coefficient less than 0.85), A2 are those who walk at 70% of the 6MWT (coefficient between 0.86 and 1.2), and A3 are those walking at 100% of the 6MWT (coefficient above 1.2) in the fifth session.

### 2.4. Data Analysis

The kinematic analysis was performed using the Nexus software [15]. Firstly, we investigate if the subjects presented any asymmetries of maximal flexion of hip, knee, and ankle, abduction ROM of hip and the maximal abduction of hip and flexion and extension maximal of knee on touch down and, stance time(s), relative stance time (%) and double stance time(s). After, we tested the effect of NW on the asymmetrical variables. The angles were determined by software VICON NEXUS®1.8, that use Euler calculations, all lower and upper body angles are calculated in rotation order YXZ, except for ankle angles, which are calculated in order YZX (available on Plug-in Gait Reference Guide).

The spatiotemporal variables were determined from the touch down and takeoff by 10 strides in the gait cycle. The mean of 10 strides was used to calculated angle and spatiotemporal outcomes. The data was exported from Vicon NEXUS® 1.8.5 software and processed in the LabVIEW software (National Instruments 8.5). The symmetry between the segments was considered when no statistical differences were observed in the parameters measured bilaterally [23].

### 2.5. Statistical Analysis

We used descriptive analysis to report the results (mean and confidence interval Wald 95%). The Generalized Estimating Equations (GEE) was used to characterize the gait asymmetries. When $p < 0.05$, the variables were considered asymmetric. Additionally, GEE was used to test the main effects of NW in asymmetrical variables, and the Bonferroni post hoc test was performed to identify the significant differences at $\alpha < 0.05$. Also, the effect size (ES) was calculated using the Hedges' g considering trivial (<0.20), small (0.20–0.49), moderate (0.50–0.79), large (>0.80), and too large (>1.30) effects between pre and post-tests at more affected and less affected segments in the asymmetrical variables [30,31]. Statistical analysis was performed by a highly trained researcher who was blinded to the participants, using SPSS software (Statistical Package for Social Sciences, version 21.0).

## 3. Results

A total of 14 participants with idiopathic PD were included in the study. Individual characteristics of the sample are shown in Table 1.

**Table 1.** Characteristics of the subjects.

| Variable | Mean (Standard Deviation) |
| --- | --- |
| Total subjects (male/female) | 14 (7/7) |
| Gender (female/male) | 7/7 |
| Age (years) | 66.8 (±9.6) |
| Disease duration (years) | 7.2 (±5.4) |
| UPDRS (points) | 12.2 (±6.1) |
| H & Y | 1.5 (1–3) |
| MoCA | 26.6 (2.2) |
| Lower limb length (m) | 0.89 (0.05) |
| Body mass (kg) | 64.5 (±23.5) |
| Height (m) | 1.7 (±.86) |

The results showed asymmetries in baseline with a significant difference for maximum knee flexion at 0.28 m·s$^{-1}$. Time-condition interaction analysis ($p = 0.007$) showed that the improvement occurred only in the less affected limb [$p < 0.001$ (ES: 0.82)] when compared to the more affected limb (Figure 2A and Table 2).

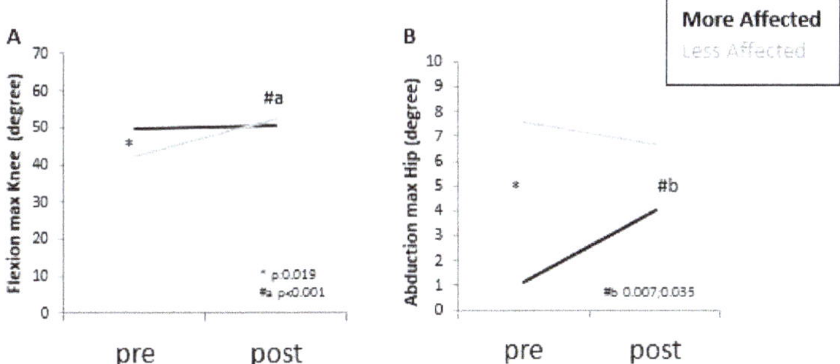


**Figure 2.** Asymmetric variables between less affected (grey line) and more affected (black line) limbs, pre and post-intervention at 0.28 m·s$^{-1}$, based in post hoc values. *: Difference between conditions on pre; (**A**) Difference between pre and post in the less affected side; (**B**) Difference between pre and post in both groups.

**Table 2.** Mean, confidence interval, minimal (min) and maximal (max) values, and effect size of Maximal angular of flexion and abduction in more and less affected segments at 0.28 m·s$^{-1}$.

| | | PRE | | POST | | | | | | |
| | | More Affected | Less Affected | More Affected | Less Affected | *p*-Value | | | | |
| | Speed (m·s$^{-1}$) | Mean (max; min) | Mean (max; min) | Mean (max; min) | Mean (max; min) | T | C | T*C | ES More Affected | ES Less Affected |
|---|---|---|---|---|---|---|---|---|---|---|
| Flexion | - | - | - | - | - | - | - | - | - | - |
| Knee (degree) | 0.28 | 49.9 (45.7; 54.1) | 42.3 (35.7; 49.0) | 50.8 (46.2; 55.4) | 52.3 (47.3; 57.3) | 0.012 * | 0.236 | 0.007 * | 0.10 | 0.82 |
| Abduction | | | | | | | | | | |
| Hip (degree) | 0.28 | 1.1 (−1.9; 4.1) | 7.6 (4.7; 10.5) | 4.00 (2.04; 5.96) | 6.7 (4.7; 8.6) | 0.329 | 0.007 * | 0.040 * | 0.56 | 0.19 |

NOTE: ES: Effect size; T: General effect of time; C: General effect of condition; T*C: Interaction between time and condition * *p*: < 0.0.

Additionally, there was asymmetry in the maximal hip abduction with time-condition interaction (Figure 2B and Table 2) at speed 0.28 m·s$^{-1}$ ($p$ = 0.040). The maximal hip abduction was increased between pre and post in both groups [$p$ = 0.035 (ES = 0.56 more affected limb)] and [$p$ = 0.007 (ES: 0.19 less affected limb)]. The other variables did not show asymmetries and are represented in supplementary material 2.

## 4. Discussion

The main purpose of this study was to characterize the gait asymmetries and then investigate the effects of NW training in the asymmetries variables. Before the intervention, the individuals with mild PD walking at 0.28 m·s$^{-1}$ showed asymmetries only in maximal knee flexion and maximal hip abduction. We refuted the hypothesis because most movements were symmetric in the baseline. On the other hand, secondary asymmetries were interestingly found at the knee and hip joints. The NW training was able to improve these parameters, becoming more symmetrical in these parameters after the intervention.

In PD, the basal ganglia dysfunction contributes to more significant gait disturbances and symptoms are directly associated with right or left cerebral hemisphere, though mechanisms responsible for this left-right coordination are not fully understood [32,33]. Additionally, PD asymmetry can be explained for a reduced number of neurons in one side of substantia nigra [34]. However, the asymmetrical side might be merely coincidental, and in the early stage, the degeneration is lower [35]. In our study, the asymmetry between the sides was considered when the valued was less than 5% in the statistic test [23]. Probably, the general symmetry observed in the pre-test seems to be explained due to mild stage (H&Y median: 1.5 points) of PD and phase of medication "ON" utilized in the present study [26,36]. Furthermore, the disease duration is 7.3 ± 5.4 years, and, at this stage of the disease, the people with PD have a similar likelihood of unilateral and bilateral motor impairments [37]. One crucial question raised by these findings is the importance of gait analysis for detecting the motor asymmetry as a screening evaluation [38,39]. In our study, we consider gait asymmetry based on the study [23]. Also, the literature describes that gait asymmetry is related to the increase in freezing [32]. However, it is necessary to acknowledge that our PD subjects did not experience freezing of gait during the walking evaluation. Therefore, the symmetry is justified on baseline condition.

Although most kinematic variables were symmetric bilaterally, the maximum knee flexion and hip abduction at 0.28 m·s$^{-1}$ were asymmetric before the intervention. After the intervention, both variables became symmetric. The literature shows that exercise can improve Parkinson's gait performance [40]. In [41], the authors observed that NW was able to increase knee power during gait, in the more and less affected side [41]. In the present study, we observed higher maximal knee flexion in the less affected size, indicating that after NW intervention, the role of compensating the impaired movement was reduced in the less affected knee during the gait cycle. We also considered that the increase in range of knee motion might be attributed to muscle regulation as a result of lower antagonist muscle stiffness, or even because of increasing muscular strength and power of the knee flexion agonist muscles. Although the main component of this increase remains unknown, greater freedom of joint movement contributes to a more symmetrical gait and a lower risk of falls.

Higher values of maximal hip abduction and symmetry between sides were found after NW intervention in our subjects. In line with these findings, [42] showed that after treadmill training, PD subjects improve the angle of hip abduction of the dominant and non-dominant sides [42]. The higher pelvic rotation could explain the higher angle of hip abduction. However, no rotational movements were measured in this study. The symmetry found in the hip abduction found out after the training period may contribute to functional mobility due to reduced energy expenditure and lower risk of falling. Also, the larger range of hip abduction motion during gait may be related to reduced muscle and joint stiffness. Although Nordic walking (NW) uses stick support, the gains acquired from training seem to benefit the free-walking pattern of subjects with PD. Besides the improvement, the maximal abduction motion observed in our study (4 to 7 degrees) is still lower than healthy people, approximately 10 degrees [6]. Another important point to highlight is that our test was measured on a treadmill, and the angular and spatiotemporal variables can be changed when compared with free walking [43]. We suggest more studies that compared gait asymmetries in these conditions, and to obtain more details, it is important to do correlations with oxygen consumption and anatomical/structural symmetry of the lower limbs [44].

The NW is an intervention where the action from upper and lower limbs is required, resulting in a synergy between upper and lower limb muscles. The poles promote more considerable support base and weight discharge [45]. These characteristics can improve the maximal ROM of the knee and hip during gait of people with PD. Therefore, the improvement in the quality of movements during gait is critical in the daily life of this population. This study has some limitations, including: (1) it did not have a control group, (2) the control regarding the physical activity quantity from the participants on baseline (very active, active, inactive, and sedentary), and (3) we did not evaluate the freezing of gait, which could have a better detection of gait asymmetry.

## 5. Conclusions

The hypothesis was not supported, and our findings demonstrate that subjects with mild PD had symmetric gait before the intervention, except for maximal hip abduction and maximal knee flexion. NW enhanced the kinematic pattern of knee and hip joints during free-walking on the treadmill, and the more affected and less affected side became symmetric. The improvement of the range of knee and hip motion is crucial to improve the functionality of subjects with PD.

**Supplementary Materials:** The following are available online at http://www.mdpi.com/2073-8994/11/12/1481/s1, TIDier items S1: Nordic Walking intervention, Table S4: Mean, confidence interval and statistical significance of maximal flexion and abduction variables on 0.28 and 0.83 m·s$^{-1}$. Table S5: Mean, confidence interval and statistical significance and range of motion of more and less affected segments on 0.28 and 0.83 m·s$^{-1}$. Table S6: Mean, confidence interval and statistical significance of spatiotemporal variables on 0.28 and 0.83 m·s$^{-1}$.

**Author Contributions:** Conception of project, A.P.J.Z. and L.A.P.-T.; Organization, A.P.J.Z. and L.A.P.-T.; Execution, A.P.J.Z., M.Z.C. and V.F.M.; Statistical analysis, A.P.J.Z. and V.F.M.; Manuscript Preparation, A.P.J.Z., F.G.M., E.S.d.S. and L.A.P.-T.; Writing of the first draft, A.P.J.Z.; Review and Critique, A.P.J.Z., F.G.M., E.S.d.S., M.Z.C., V.F.M., E.P.-M., A.N.H. and L.A.P.-T.

**Funding:** This research was funded by Coordenação de Aperfeiçoamento de Pessoal de Nível Superior—Brasil (CAPES)—Finance Code 001.

**Conflicts of Interest:** The authors declare that they have no conflicts of interest relevant to the content of this review.

## References

1. Morris, S.; Morris, M.E.; Iansek, R. Reliability of measurements obtained with the Timed "Up & Go" test in people with Parkinson disease. *Phys. Ther.* **2001**, *81*, 810–818. [CrossRef] [PubMed]
2. Frazzitta, G.; Pezzoli, G.; Bertotti, G.; Maestri, R. Asymmetry and freezing of gait in parkinsonian patients. *J. Neurol.* **2013**, *260*, 71–76. [CrossRef] [PubMed]
3. Williams, A.J.; Peterson, D.S.; Earhart, G.M. Gait coordination in Parkinson disease: Effects of step length and cadence manipulations. *Gait Posture* **2013**, *38*, 340–344. [CrossRef] [PubMed]
4. Martinez, M.; Villagra, F.; Castellote, J.M.; Pastor, M.A. Kinematic and Kinetic Patterns Related to Free-Walking in Parkinson's Disease. *Sensors* **2018**, *18*, 4224. [CrossRef]
5. Boonstra, T.A.; van der Kooij, H.; Munneke, M.; Bloem, B.R. Gait disorders and balance disturbances in Parkinson's disease: Clinical update and pathophysiology. *Curr. Opin. Neurol.* **2008**, *21*, 461–471. [CrossRef]
6. Saunders, J.B.D.M.; Inman, V.T.; Eberhart, H.D. The major determinants in normal and pathological gait. *J. Bone Jt. Surg. Am.* **1953**, *35*, 543–558. [CrossRef]
7. Cavagna, A.; Thys, H.; Zamboni, A. The sources of external work in level walking and running. *J. Physiol.* **1976**, *262*, 639–657. [CrossRef]
8. Bianchi, L.; Angelini, D.; Orani, G.P.; Lacquaniti, F. Kinematic coordination in human gait: Relation to mechanical energy cost. *J. Neurophysiol.* **1998**, *79*, 2155–2170. [CrossRef]

9.  Goodwin, V.A.; Richards, S.H.; Taylor, R.S.; Taylor, A.H.; Campbell, J.L. The effectiveness of exercise interventions for people with Parkinson's disease: A systematic review and meta-analysis. *Mov. Disord.* **2008**, *23*, 631–640. [CrossRef]

10. Shu, H.F.; Yang, T.; Yu, S.X.; Huang, H.D.; Jiang, L.L.; Gu, J.W.; Kuang, Y.Q. Aerobic exercise for Parkinson's disease: A systematic review and meta-analysis of randomized controlled trials. *PLoS ONE* **2014**, *9*, e100503. [CrossRef]

11. Hubble, R.P.; Naughton, G.; Silburn, P.A.; Cole, M.H. Trunk Exercises Improve Gait Symmetry in Parkinson Disease: A Blind Phase II Randomized Controlled Trial. *Am. J. Phys. Med. Rehabil.* **2018**, *97*, 151–159. [CrossRef] [PubMed]

12. Cugusi, L.; Manca, A.; Dragone, D.; Deriu, F.; Solla, P.; Secci, C.; Monticone, M.; Mercuro, G. Nordic Walking for the Management of People With Parkinson Disease: A Systematic Review. *PM&R* **2017**, *9*, 1157–1166. [CrossRef]

13. Monteiro, E.P.; Franzoni, L.T.; Cubillos, D.M.; de Oliveira Fagundes, A.; Carvalho, A.R.; Oliveira, H.B.; Pantoja, P.D.; Schuch, F.B.; Rieder, C.R.; Martinez, F.G.; et al. Effects of Nordic walking training on functional parameters in Parkinson's disease: A randomized controlled clinical trial. *Scand. J. Med. Sci. Sports* **2017**, *27*, 351–358. [CrossRef] [PubMed]

14. Franzoni, L.T.; Monteiro, E.P.; Oliveira, H.B.; da Rosa, R.G.; Costa, R.R.; Rieder, C.; Martinez, F.G.; Peyré-Tartaruga, L.A. A 9-week Nordic and free walking improve postural balance in Parkinson's disease. *Sports Med. Int. Open* **2018**, *2*, 28–34. [CrossRef] [PubMed]

15. Gomeñuka, N.A.; Oliveira, H.B.; Silva, E.S.; Costa, R.R.; Kanitz, A.C.; Liedtke, G.V.; Schuch, F.B.; Peyré-Tartaruga, L.A. Effects of Nordic walking training on quality of life, balance and functional mobility in elderly: A randomized clinical trial. *PLoS ONE* **2019**, *14*, e0211472. [CrossRef] [PubMed]

16. Arcila, D.M.C.; Monteiro, E.P.; Gomeñuka, N.A.; Peyré-Tartaruga, L.A. Metodologia e Didática Pedagógica aplicada ao ensino da Caminhada Nórdica e Livre para pessoas com Doença de Parkinson I [Methodology and pedagogical didactics applied to the education of nordic walking and free walking for people with parkinson's disease I]. *Cad. RBCE* **2018**, *8*, 72–83.

17. Pellegrini, B.; Peyré-Tartaruga, L.A.; Zoppirolli, C.; Bortolan, L.; Savoldelli, A.; Minetti, A.E.; Schena, F. Mechanical energy patterns in Nordic walking: Comparisons with conventional walking. *Gait Posture* **2017**, *51*, 234–238. [CrossRef]

18. Boccia, G.; Zoppirolli, C.; Bortolan, L.; Schena, F.; Pellegrini, B. Shared and task-specific muscle synergies of Nordic walking and conventional walking. *Scand. J. Med. Sci. Sports* **2018**, *28*, 905–918. [CrossRef]

19. Pellegrini, B.; Peyré-Tartaruga, L.A.; Zoppirolli, C.; Bortolan, L.; Bacchi, E.; Figard-Fabre, H.; Schena, F. Exploring muscle activation during Nordic walking: A comparison between conventional and uphill walking. *PLoS ONE* **2015**, *10*, e0138906. [CrossRef]

20. Miller, R.A.; Thaut, M.H.; McIntosh, G.C.; Rice, R.R. Components of EMG symmetry and variability in parkinsonian and healthy elderly gait. *Electroencephalogr. Clin. Neurophysiol.* **1996**, *101*, 1–7. [CrossRef]

21. Morris, M.E.; Huxham, F.; McGinley, J.; Dodd, K.; Iansek, R. The biomechanics and motor control of gait in Parkinson disease. *Clin. Biomech.* **2001**, *16*, 459–470. [CrossRef]

22. Delval, A.; Salleron, J.; Bourriez, J.L.; Bleuse, S.; Moreau, C.; Krystkowiak, P.; Defebvre, L.; Devos, P.; Duhamel, A. Kinematic angular parameters in PD: Reliability of joint angle curves and comparison with healthy subjects. *Gait Posture* **2008**, *28*, 495–501. [CrossRef] [PubMed]

23. Sadeghi, H.; Allard, P.; Prince, F.; Labelle, H. Symmetry and limb dominance in able-bodied gait: A review. *Gait Posture* **2000**, *12*, 34–45. [CrossRef]

24. Mehrholz, J.; Kugler, J.; Storch, A.; Pohl, M.; Hirsch, K.; Elsner, B. Treadmill training for patients with Parkinson Disease. An abridged version of a Cochrane Review. *Eur. J. Phys. Rehabil. Med.* **2016**, *52*, 704–713. [PubMed]

25. Tumas, V.; Borges, V.; Ballalai-Ferraz, H.; Zabetian, C.P.; Mata, I.F.; Brito, M.M.C.; Foss, M.P.; Novaretti, N.; Santos-Lobato, B.L. Some aspects of the validity of the Montreal Cognitive Assessment (MoCA) for evaluating cognitive impairment in Brazilian patients with Parkinson's disease. *Dement. Neuropsychol.* **2016**, *10*, 333–338. [CrossRef] [PubMed]

26. Ribeiro, T.S.; de Sousa, A.C.; de Lucena, L.C.; Santiago, L.M.M.; Lindquist, A.R.R. Does dual task walking affect gait symmetry in individuals with Parkinson's disease? *Eur. J. Physiother.* **2018**, *21*, 8–14. [CrossRef]

27. O'neal, H.A.; Blair, S.N. Enhancing Adherence in Clinical Exercise Trials. *Quest* **2001**, *53*, 310–317. [CrossRef]

28. Raciti, L.; Nicoletti, A.; Mostile, G.; Bonomo, R.; Contrafatto, D.; Dibilio, V.; Luca, A.; Sciacca, G.; Cicero, C.E.; Vasta, R.; et al. Validation of the UPDRS section IV for detection of motor fluctuations in Parkinson's disease. *Park. Relat. Disord.* **2016**, *27*, 98–101. [CrossRef]

29. Enright, P.L.; McBurnie, M.A.; Bittner, V.; Tracy, R.P.; McNamara, R.; Arnold, A.; Newman, A.B. The 6-min walk test: A quick measure of functional status in elderly adults. *Chest* **2003**, *123*, 387–398. [CrossRef]

30. Rosenthal, J.A. Qualitative Descriptors of Strength of Association and Effect Size. *J. Soc. Serv. Res.* **1996**, *21*, 37–59. [CrossRef]

31. Espirito-santo, H.; Daniel, F. Calcular e apresentar tamanhos do efeito em trabalhos científicos (1): As limitações do $p < 0.05$ na análise de diferenças de médias de dois grupos. *Rev. Port. Invest. Comp. Soc.* **2015**, *1*, 3–6.

32. Plotnik, M.; Giladi, N.; Balash, Y.; Peretz, C.; Hausdorff, J.M. Is freezing of gait in Parkinson's disease related to asymmetric motor function? *Ann. Neurol.* **2005**, *57*, 656–663. [CrossRef] [PubMed]

33. Lee, E.Y.; Sen, S.; Eslinger, P.J.; Wagner, D.; Kong, L.; Lewis, M.M.; Du, G.; Huang, X. Side of motor onset is associated with hemisphere-specific memory decline and lateralized gray matter loss in Parkinson's disease. *Park. Relat Disord* **2015**, *21*, 465–470. [CrossRef] [PubMed]

34. Djaldetti, R.; Ziv, I.; Melamed, E. The mystery of motor asymmetry in Parkinson's disease. *Lancet Neurol.* **2006**, *5*, 796–802. [CrossRef]

35. Mirelman, A.; Bonato, P.; Camicioli, R.; Ellis, T.D.; Giladi, N.; Hamilton, J.L.; Hass, C.J.; Hausdorff, J.M.; Pelosin, E.; Almeida, Q.J. Gait impairments in Parkinson's disease. *Lancet Neurol.* **2019**, *18*, 697–708. [CrossRef]

36. Yogev, G.; Plotnik, M.; Peretz, C.; Giladi, N.; Hausdorff, J.M. Gait asymmetry in patients with Parkinson's disease and elderly fallers: When does the bilateral coordination of gait require attention? *Exp. Brain Res.* **2007**, *177*, 336–346. [CrossRef]

37. Schenkman, M.L.; Clark, K.; Xie, T.; Kuchibhatla, M.; Shinberg, M.; Ray, L. Spinal movement and performance of a standing reach task in participants with and without Parkinson disease. *Phys. Ther.* **2001**, *81*, 1400–1411. [CrossRef]

38. Djurić-Jovičić, M.; Belić, M.; Stanković, I.; Radovanović, S.; Kostić, V.S. Selection of gait parameters for differential diagnostics of patients with de novo Parkinson's disease. *Neurol. Res.* **2017**, *39*, 853–861. [CrossRef]

39. Meyer, C.; Killeen, T.; Easthope, C.S.; Curt, A.; Bolliger, M.; Linnebank, M.; Zörner, B.; Filli, L. Familiarization with treadmill walking: How much is enough? *Sci. Rep.* **2019**, *9*, 5232. [CrossRef]

40. Ni, M.; Hazzard, J.B.; Signorile, J.F.; Luca, C. Exercise Guidelines for Gait Function in Parkinson's Disease: A Systematic Review and Meta-analysis. *Neurorehabil. Neural Repair* **2018**, *32*, 872–886. [CrossRef]

41. Zhou, L.; Gougeon, M.A.; Nantel, J. Nordic Walking Improves Gait Power Profiles at the Knee Joint in Parkinson's Disease. *J. Aging Phys. Act.* **2018**, *26*, 84–88. [CrossRef] [PubMed]

42. Luna, N.M.S.; Lucareli, P.R.G.; Sales, V.C.; Speciali, D.; Alonso, A.C.; Peterson, M.D.; Rodrigues, R.B.M.; Fonoffc, E.T.; Barbosac, E.R.; Teixeira, M.J.; et al. Treadmill training in Parkinson's patients after deep brain stimulation: Effects on gait kinematic. *Neurorehabilitation* **2018**, *42*, 149–158. [CrossRef] [PubMed]

43. Malatesta, D.; Canepa, M.; Fernandez, A.M. The effect of treadmill and overground walking on preferred walking speed and gait kinematics in healthy, physically active older adults. *Eur. J. Appl. Physiol.* **2017**, *117*, 1833–1843. [CrossRef] [PubMed]

44. Seminati, E.; Nardello, F.; Zamparo, P.; Ardigó, L.P.; Faccioli, N.; Minetti, A.E. Anatomically Asymmetrical Runners Move More Asymmetrically at the Same Metabolic Cost. *PLoS ONE* **2013**, *8*, e74134. [CrossRef] [PubMed]

45. Dziuba, A.K.; Żurek, G.; Garrard, I.; Wierzbicka-Damska, I. Parameters in lower limbs during natural walking and Nordic walking at different speeds. *Acta Bioeng. Biomech.* **2015**, *17*, 95–101. [PubMed]

Article

# The Effect of a Secondary Task on Kinematics during Turning in Parkinson's Disease with Mild to Moderate Impairment

Francesca Nardello, Emanuele Bertoli, Federica Bombieri, Matteo Bertucco * and Andrea Monte

Department of Neurosciences, Biomedicine and Movement Sciences University of Verona, 43, 37131 Verona, Italy; francesca.nardello@univr.it (F.N.); emanuele.bertoli@studenti.univr.it (E.B.); federica.bombieri@univr.it (F.B.); andrea.monte@univr.it (A.M.)
* Correspondence: matteo.bertucco@univr.it; Tel.: +39-045-8425131

Received: 10 July 2020; Accepted: 29 July 2020; Published: 3 August 2020

**Abstract:** Patients with Parkinson's disease (PD) show typical gait asymmetries. These peculiar motor impairments are exacerbated by added cognitive and/or mechanical loading. However, there is scarce literature that chains these two stimuli. The aim of this study was to investigate the combined effects of a dual task (cognitive task) and turning (mechanical task) on the spatiotemporal parameters in mild to moderate PD. Participants (nine patients with PD and nine controls (CRs)) were evaluated while walking at their self-selected pace without a secondary task (single task), and while repeating the days of the week backwards (dual task) along a straight direction and a 60° and 120° turn. As speculated, in single tasking, PD patients preferred to walk with a shorter stride length ($p < 0.05$) but similar timing parameters, compared to the CR group; in dual tasking, both groups walked slower with shorter strides. As the turn angle increased, the speed will be reduced ($p < 0.001$), whereas the ground–foot contact will become greater ($p < 0.001$) in all the participants. We showed that the combination of a simple cognitive task and a mechanical task (especially at larger angles) could represent an important training stimulus in PD at the early stages of the pathology.

**Keywords:** Parkinson's disease; gait; dual task; turning; kinematics

## 1. Introduction

Gait functionality is accomplished with good health, whereas locomotion disturbances are associated with an advancing age or with the presence of pathologies. Within 3 years of diagnosis, over 85% of people with Parkinson's disease (PD) develop gait deteriorations [1,2], as well as a reduced walking symmetry [2–4]. Such motor dysfunctions could be exacerbated under various conditions, such as the dual task performance and the changing of the walking direction (i.e., turning).

Specifically, the execution of the two tasks simultaneously ("concurrent performance") presents a challenge for PD patients because of their disabled lower-level spinal centers and basal nuclei [5–7]. Studies have shown that, when compared to controls, subjects with PD show greater reductions in their stride length [8,9] and, therefore, walking speed, as well as an increased variability of the kinematic parameters [10–16]. These patterns are aggravated with the severity of the pathology [17,18] and with the complexity of the concurrent task [5,13,19,20]. Since these findings strengthen the idea that the cognitive function could contribute to gait regulation, it could be helpful to understand the effect of different cognitive tasks, even very simple ones, in this kind of pathology [21–23]. This knowledge could provide an insight on what is the best training stimulus to be administered in those patients.

Concerning the walking direction, it has been shown that subjects with PD report turning difficulties, which are associated with an augmented risk of falling [24–27]. These impairments are due to the central nervous system's involvement in body re-orientation when travelling in the new direction.

From a kinematical point of view, patients with PD highlighted a smaller step width [9,28], a reduction in the step length and a decrease in their walking velocity during a change of direction of 45° in the stepping tasks [29]. If the turning angle increased (i.e., to 90°), patients approached turns with a slower step length and by performing the turn with a larger number of steps [30]. Walking with an auditory cue reduced the gait-timing variability, as well as the step length, and increased the radius of gyration during a turn of 180° [31]. Moreover, recent literature has shown that subjects with PD spent more time during the turn (between 30° and 180°) and exhibit a reduction in the walking stability compared to controls [32]. Turning 360° in place seemed to be a more compromised condition when compared to turning while walking, with similar impairments with and without episodes of freezing [33,34]. As in the case of the cognitive additional demand, the turning conditions could also represent an important stimulus in such a population. Hence, investigating different turning conditions could provide important information about the optimal training strategy in people with PD.

To date, a lingering question in the literature regards the combination of mechanical and cognitive perturbations. Indeed, few studies have combined the dual task condition with the mechanical perturbations (e.g., change of direction) [35–37]. Furthermore, previous studies usually used turn angles larger than 90°, which represents per se an important perturbation for PD patients [30].

Therefore, we used the most common turning angles during daily life (60° and 120°) [38], in combination with a simple cognitive task (repeating the days of the week backwards), to better understand the effects of these perturbations on gait kinematics. This combined approach has involved participants with mild to moderate PD in order to provide additional information on how the typical Parkinson's disease walking pattern is modified under simultaneous cognitive and mechanical loads.

Hence, the primary aim was to investigate if the presence of a "dual task" condition can further alter the kinematics (i.e., spatiotemporal parameters) of walking, or, on the contrary, if it can represent a good training stimulus in mild and moderate Parkinson's disease. This cognitive task was studied during forward/linear walking and during turning.

Based on the previously mentioned literature, we hypothesized that the simple cognitive and the mechanical tasks will exhibit similar effects in both populations (patients and controls), whereas the combination of the two stimuli will show a higher impact on PD patients.

## 2. Materials and Methods

### 2.1. Participants

This study has a cross-sectional, analytical, observational design. Nine patients with mild to moderate PD (3 women and 6 men; mean ± SD, 68.2 ± 5.95 years old, 74.2 ± 11.8 kg body mass, 1.70 ± 0.10 m height, and 25.6 ± 3.07 kg·m$^{-2}$ body mass index (BMI)) and nine healthy age-matched controls (4 women and 5 men; 67.2 ± 3.45 years old, 72.6 ± 13.1 kg, 1.68 ± 0.08 m, and 25.7 ± 3.43 kg·m$^{-2}$) were recruited in this study.

All participants received written and oral instructions before the study and gave their written informed consent for the experimental procedure. The study was conducted in accordance with the Declaration of Helsinki, and the experimental protocol was approved by the Institutional Review Board of the Department of Neurosciences, Biomedicine and Movement Sciences, University of Verona (protocol number 2018-UNVRCLE-0451799).

Participants were recruited from a sample of late adulthood people attending the adapted physical activity program at the School of Sport and Exercise Science at our University (see their characteristics in Table 1).

**Table 1.** Characteristics of subjects with Parkinson's disease.

| Subject No. | Age (yrs) | Sex | Body Mass (kg) | Body Height (m) | BMI (kg·m$^{-2}$) | Disease Duration (yrs) | Y&H Score | UPDRS Score | MMSE | Medication | Side of Appearance of the First Motor Symptom |
|---|---|---|---|---|---|---|---|---|---|---|---|
| PD1 | 77 | F | 80 | 1.64 | 29.74 | 4 | 1 | 26 | 29 | Levodopa/carbidopa | Left |
| PD2 | 66 | F | 63 | 1.58 | 25.24 | 4 | 3 | 37 | 27 | Levodopa/carbidopa | Left |
| PD3 | 71 | M | 70 | 1.70 | 24.22 | 3 | 1 | 24 | 28 | Levodopa/benserazide | Right |
| PD4 | 59 | M | 65 | 1.70 | 22.49 | 9 | 1 | 26 | 29 | Levodopa/carbidopa | Left |
| PD5 | 67 | M | 80 | 1.66 | 29.03 | 3 | 1 | 30 | 26 | Levodopa/carbidopa | Right |
| PD6 | 67 | M | 80 | 1.77 | 25.54 | 6 | 1 | 30 | 28 | Levodopa/benserazide | Left |
| PD7 | 63 | F | 55 | 1.59 | 21.76 | 1 | 1 | 23 | 29 | Levodopa/benserazide | Right |
| PD8 | 77 | M | 82 | 1.88 | 23.20 | 8 | 1 | 25 | 24 | Levodopa/carbidopa | Right |
| PD9 | 67 | M | 93 | 1.78 | 29.35 | 7 | 2 | 32 | 25 | Levodopa/carbidopa | Left |

BMI: body mass index; Y&H: Hoehn and Yahr; UPDRS: Unified Parkinson's Disease Rating Scale (motor section); MMSE: Mini Mental State Education.

Patients were excluded if their medical condition proved unstable due to other neurological, orthopedic, metabolic or cardiovascular co-morbidity factors affecting gait. There was no type of rehabilitation in the month prior to recruitment or disease-modifying therapy that was not well defined. A diagnosis of idiopathic PD was carried out by a neurologist, in accordance with the guidelines established by the London Brain Bank [39]. The disease severity was classified according to the modified Hoehn and Yahr scale (H&Y) [40], whereas the assessment of the degree of motor and functional impairment was obtained using part III of the Unified Parkinson's Disease Rating Scale (UPDRS) [41]. Finally, cognitive function was assessed by using the Mini Mental State Examination (MMSE), with scores of up to 30 (higher scores correspond to greater cognitive function).

We conducted an a priori screening to evaluate whether participants were able to carry out the task instructions required for the experimental protocol (see the "Experimental procedures" later). Based on this a priori screening, only individuals with PD with an MMSE score of 24 or higher were able to perform the double stimuli, and therefore were recruited in the study.

### 2.2. Experimental Procedures

Tests were conducted in the Biomechanics Laboratory at the Department of Neurosciences, Biomedicine and Movement Sciences. The spatial (distance) and temporal (time) characteristics of the step pattern were measured using an eight-camera motion capture system (MX Ultranet, VICON, Oxfordshire, UK), sampling at 100 Hz. This apparatus recorded the position of markers positioned bilaterally on the feet: the calcaneus, lateral malleolus and 5th metatarsal-phalangeal joint.

Participants started from a standing position and walked barefoot at their self-selected speed over a 10 m walkway (Figure 1, panel a). Each participant performed three different conditions: *(i)* forward walking (WF (Figure 1, panel b)); *(ii)* walking, turning at 60°, walking (T60); *(iii)* walking, turning at 120°, walking (T120). Furthermore, T60 and T120 were conducted turning in both directions (left and right (Figure 1, panels c and d)). During all turning tasks, a cone was positioned in the center of the walkway (see the black hexagon in Figure 1) to identify the turning point. Furthermore, an operator positioned himself at the end of the corridor lane that the subjects had to move. All directions were tested in single and dual tasks, and the cognitive load consists of walking while repeating the days of the week backwards [17]. Six trials of each condition were performed and the order of gait conditions was randomly allocated (Figure 2). Subjects sat and rested in a chair for three minutes between trials. Patients were tested at the peak dose in the medication cycle.

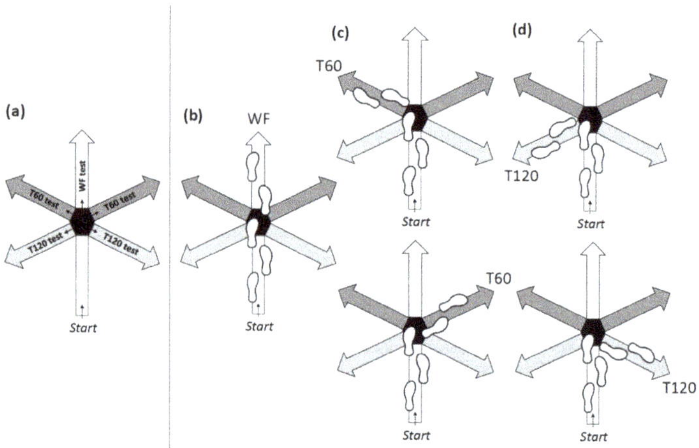


**Figure 1.** Schematic representation of walking conditions (**a**), and steps during walking (forward walking (WF), (**b**) and turning (T60 in (**c**) and T120 in (**d**)).

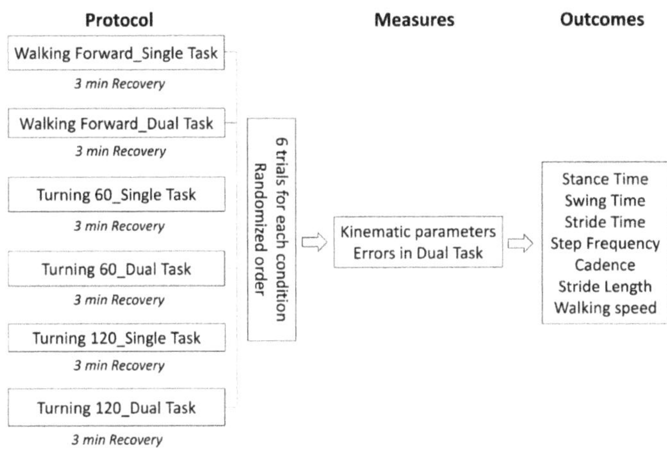



**Figure 2.** Flow chart of the research design.

### 2.3. Data Reduction

Kinematic data were recorded from 3 m before and after the turning step. The quantitative gait assessment included both temporal and spatial parameters (Figure 3), that are: *(i)* stance time (s), the period of time when the foot is in contact with the ground; *(ii)* swing time (s), the period of time when the foot is not in contact with the ground; *(iii)* stride time (s), the interval of time to complete a gait stride; *(iv)* cadence (stride·min$^{-1}$), the number of strides taken in a unit of time; *(v)* stride length (m), the distance from the initial contact of one foot to the following initial contact of the same foot. All these parameters were obtained using standard definitions according to an algorithm programmed in LabView (version 10, National Instruments, Austin, TX, USA). Walking speed (m·s$^{-1}$) was appreciated as the distance travelled during a complete stride cycle.

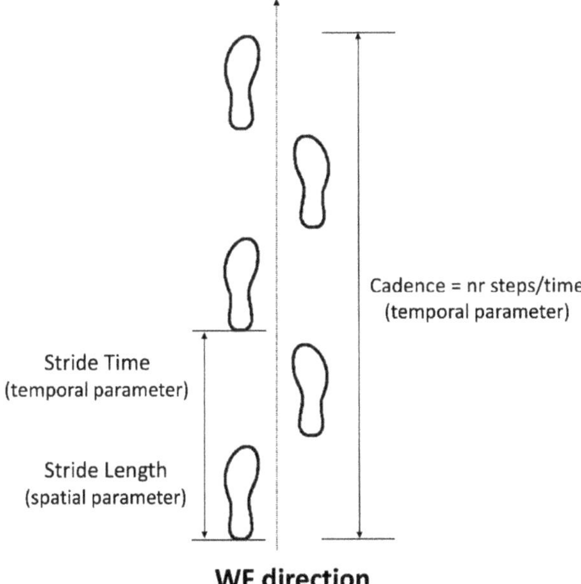



**Figure 3.** Gait parameter assessment for walking/turning conditions.

The average value of all the performed steps in each condition has been utilized in further analyses. The number of errors (i.e., when telling a day backwards went wrong) was counted and collected by an operator during the walking trials as well. For controls, we pooled together the data from the right and left lower limbs ($n = 9$), whereas for PD patients, the data referring to the less affected side (PDNA) and to the more affected side (PDA) were considered separately ($n = 9$).

Finally, we calculated the locomotor rehabilitation index "LRI" as: LRI = 100 * (self-selected speed (SSWS)/optimal speed (OWS)), where the self-selected speed (SSWS) has been directly measured and the optimal speed (OWS) has been estimated according to a previous study [42].

## 2.4. Data Analysis

The data analysis was conducted using SPSS 19.0 (SPSS Inc, Chicago, IL, USA). Descriptive statistics were used to compute the means and standard deviations for the outcome variables. All the spatiotemporal gait data were normally distributed (Kolmogorov–Smirnov and Shapiro–Wilk tests) and did not violate the assumptions of homogeneity. In order to test the main hypotheses, a series of 2 (tasks: single task vs. dual task) × 3 (gait directions: WF vs. T60 vs. T120) repeated measures ANOVAs, with groups (CR, PDNA, PDA) as the between factors, were used to analyze the gait data. When a significant main effect was found (critical $p$ value $< 0.05$), a post-hoc $t$-test was performed. A Bonferroni correction was applied when needed.

## 3. Results

### 3.1. Cognitive Task

For the MMSE, controls ($28.1 \pm 1.66$) showed similar scores to patients ($27.2 \pm 1.85$ (p = ns)).

The majority of the dual task trails were successful. Whereas individuals with PD performed the cognitive task with $97.2 \pm 6\%$, $94.9 \pm 8\%$ and $95.2 \pm 10\%$ accuracy during WF, T60 and T120, respectively, and controls performed the same trails with an accuracy of $96.7 \pm 5\%$, $95.3 \pm 6\%$ and $94.9 \pm 9\%$. Regarding the gait direction, more correct answers were given while walking forward as compared to turning ($p < 0.01$ for T60 and T120) similarly in both groups.

### 3.2. Spatiotemporal Parameters

There were no differences between groups for age, body mass and height, or body mass index.

The average $\pm$ SD values of the measured temporal and spatial variables have been reported in Tables 2 and 3, as well as the post-hoc results.

The stance time differed among tasks (F (1, 24) = 34.456, $p < 0.001$) and walking direction (F (2, 48) = 27.234, $p < 0.001$), whereas there were no differences between groups (F (2, 24) = 0.262, p = ns). In particular, the dual task showed an increased stance time compared to the single one. The highest stance phase was observed during turning at 120°, while the lowest was during walking in the straight direction.

The swing phase showed a significant effect for direction (F (2, 48) = 4.648, $p < 0.05$), but not for task (F (1, 24) = 3.745, p = ns) and group (F (2, 24) = 0.883, p = ns). Particularly, the 120° turn took a longer swing time in comparison to the other directions.

The cycle time showed significant differences among tasks (F (1, 24) = 31.840, $p < 0.001$) and directions (F (2, 48) = 34.227, $p < 0.001$), whereas there were no group changes (F (2, 24) = 0.618, p = ns). Both groups increased the cycle time during the dual task condition, and the 120° turn took the longest time compared with the other conditions.

Therefore, the step frequency and cadence changed significantly with regard to task (F (1, 24) = 30.257 and 30.317, $p < 0.001$) and direction (F (2, 48) = 31.329 and 31.441, $p < 0.001$), but not with group (F (2, 24) = 0.458 and 0.457, p = ns).

**Table 2.** Gait temporal parameters at single and dual tasking in all directions. Data are means ± SD.

| | Stance Time (s) | | | Swing Time (s) | | |
|---|---|---|---|---|---|---|
| **Walking Direction** | **CR** | **PDNA** | **PDA** | **CR** | **PDNA** | **PDA** |
| WF_ST | 0.72 ± 0.05 | 0.74 ± 0.12 [a] | 0.74 ± 0.18 [c,l] | 0.33 ± 0.03 | 0.33 ± 0.04 | 0.33 ± 0.02 [a] |
| T60_ST | 0.73 ± 0.07 [a] | 0.76 ± 0.13 [a,h,m] | 0.75 ± 0.12 [b,g] | 0.33 ± 0.04 | 0.33 ± 0.04 | 0.35 ± 0.03 |
| T120_ST | 0.75 ± 0.07 [b] | 0.78 ± 0.14 [b,h,m] | 0.78 ± 0.14 [c,l,g] | 0.33 ± 0.04 | 0.35 ± 0.03 | 0.35 ± 0.03 |
| WF_DT | 0.75 ± 0.07 [m] | 0.80 ± 0.12 [a,m] | 0.78 ± 0.11 [c,l] | 0.33 ± 0.04 | 0.34 ± 0.04 | 0.35 ± 0.05 [a] |
| T60_DT | 0.78 ± 0.11 [a] | 0.84 ± 0.14 [a] | 0.82 ± 0.14 [b] | 0.33 ± 0.03 | 0.34 ± 0.04 | 0.35 ± 0.04 |
| T120_DT | 0.81 ± 0.11 [b,m] | 0.85 ± 0.14 [b,m] | 0.83 ± 0.13 [c,l] | 0.34 ± 0.04 | 0.35 ± 0.05 | 0.37 ± 0.06 |

| | Stride Time (s) | | | Cadence (Stride·min$^{-1}$) | | |
|---|---|---|---|---|---|---|
| **Walking Direction** | **CR** | **PDNA** | **PDA** | **CR** | **PDNA** | **PDA** |
| WF_ST | 1.05 ± 0.08 [a] | 1.07 ± 0.12 [a,n] | 1.08 ± 0.12 [b,m] | 114.79 ± 8.68 [a] | 113.55 ± 12.27 [c,n] | 112.74 ± 11.91 [b,d,m] |
| T60_ST | 1.06 ± 0.11 [a] | 1.09 ± 0.12 [a,h] | 1.10 ± 0.12 [b] | 113.77 ± 10.01 [a] | 111.35 ± 12.10 [c,i] | 109.92 ± 11.32 [b,d] |
| T120_ST | 1.08 ± 0.11 [a] | 1.14 ± 0.14 [c,n,h] | 1.13 ± 0.13 [a,m] | 112.16 ± 9.96 [b] | 106.90 ± 12.98 [a,i,n] | 107.04 ± 12.10 [a,m] |
| WF_DT | 1.09 ± 0.10 [a,l] | 1.13 ± 0.11 [a,n] | 1.13 ± 0.12 [b,n] | 110.94 ± 10.08 [a,m] | 106.76 ± 10.79 [c,n] | 106.79 ± 10.92 [b,m] |
| T60_DT | 1.11 ± 0.13 [a] | 1.18 ± 0.13 [a] | 1.18 ± 0.13 [b] | 108.88 ± 12.57 [a,g] | 102.94 ± 11.55 [c] | 103.04 ± 11.02 [b] |
| T120_DT | 1.14 ± 0.12 [a,l] | 1.20 ± 0.15 [c,n] | 1.20 ± 0.14 [a,n] | 105.84 ± 10.87 [b,g,m] | 101.00 ± 12.46 [a,n] | 101.44 ± 11.97 [a,m] |

CR: control group; PDNA: Parkinson's disease not affected side; PDA: Parkinson's disease more affected side. WF: walking forward; T60: turning at 60°; T120: turning at 120°. ST: single task; DT: dual task. [a] = $p < 0.05$, [b] = $p < 0.01$, [c] = $p < 0.001$: significant difference by comparing ST to DT. [d] = $p < 0.05$: significant difference by comparing the WF to T60. [g] = $p < 0.05$; [h] = $p < 0.01$, [i] = $p < 0.001$: significant difference by comparing the T60 to T120. [l] = $p < 0.05$, [m] = $p < 0.01$, [n] = $p < 0.001$: significant difference by comparing the WF to T120.

**Table 3.** Step frequency, stride length and walking speed at single and dual tasking in all directions. Data are means ± SD.

| | Step Frequency (Hz) | | | Stride Length (m) | | | Walking Speed (m·s$^{-1}$) | | |
|---|---|---|---|---|---|---|---|---|---|
| **Walking Direction** | **CR** | **PDNA** | **PDA** | **CR** | **PDNA** | **PDA** | **CR** | **PDNA** | **PDA** |
| WF_ST | 0.95 ± 0.08 [a] | 0.95 ± 0.10 [c,n] | 0.94 ± 0.10 [b,m] | 1.34 ± 0.10 [a,n,#,°] | 1.23 ± 0.14 [b,n,#] | 1.24 ± 0.14 [c,n,°] | 1.28 ± 0.12 [c,m,#,°] | 1.17 ± 0.21 [b,f,n,#] | 1.17 ± 0.21 [b,e,n,°] |
| T60_ST | 0.95 ± 0.08 [a] | 0.93 ± 0.10 [c,h] | 0.92 ± 0.10 [b] | 1.33 ± 0.08 [b,h,#,°] | 1.20 ± 0.14 [b,i,#] | 1.23 ± 0.16 [c,i,°] | 1.26 ± 0.12 [c,i,#,°] | 1.06 ± 0.17 [f,i,#] | 1.06 ± 0.18 [e,i,°] |
| T120_ST | 0.93 ± 0.08 [b] | 0.89 ± 0.11 [a,h,n] | 0.89 ± 0.10 [a,m] | 1.24 ± 0.07 [a,h,n,#,°] | 1.05 ± 0.13 [i,n,#] | 1.05 ± 0.11 [i,n,°] | 1.15 ± 0.12 [b,i,m,#,°] | 0.95 ± 0.15 [i,n,#] | 0.95 ± 0.14 [i,n,°] |
| WF_DT | 0.92 ± 0.08 [a,m] | 0.89 ± 0.09 [c,n] | 0.89 ± 0.09 [b,m] | 1.27 ± 0.06 [a,n,#,°] | 1.14 ± 0.12 [b,n,#] | 1.15 ± 0.13 [c,n,°] | 1.18 ± 0.10 [c,n,#,°] | 1.09 ± 0.17 [b,d,n,#] | 1.11 ± 0.18 [b,n,°] |
| T60_DT | 0.91 ± 0.10 [a,g] | 0.86 ± 0.10 [c] | 0.86 ± 0.09 [b] | 1.26 ± 0.07 [b,i,#,°] | 1.14 ± 0.15 [b,i,#] | 1.14 ± 0.15 [c,i,°] | 1.14 ± 0.15 [c,i,#,°] | 1.03 ± 0.18 [d,i,#] | 1.06 ± 0.17 [i,°] |
| T120_DT | 0.88 ± 0.09 [b,g,m] | 0.84 ± 0.10 [a,n] | 0.84 ± 0.10 [a,m] | 1.18 ± 0.08 [a,i,n,#,°] | 1.02 ± 0.16 [i,n,#] | 1.01 ± 0.16 [i,n,°] | 1.04 ± 0.14 [b,i,n,#,°] | 0.89 ± 0.16 [i,n,#] | 0.89 ± 0.14 [i,n,°] |

CR: control group; PDNA: Parkinson's disease not affected side; PDA: Parkinson's disease more affected side. WF: walking forward; T60: turning at 60°; T120: turning at 120°. ST: single task; DT: dual task. [a] = $p < 0.05$, [b] = $p < 0.01$, [c] = $p < 0.001$: significant difference by comparing ST to DT. [d] = $p < 0.05$, [e] = $p < 0.01$, [f] = $p < 0.001$: significant difference by comparing the WF to T60. [g] = $p < 0.05$; [h] = $p < 0.01$, [i] = $p < 0.001$: significant difference by comparing the T60 to T120. [m] = $p < 0.01$, [n] = $p < 0.001$: significant difference by comparing the WF to T120. [#] = $p < 0.05$: significant difference by comparing the CR group to the PDNA group. [°] = $p < 0.05$: significant difference by comparing the CR group to the PDA group.

The stride length was affected by group (F (2, 24) = 4.415, $p < 0.05$), task (F (1, 24) = 41.213, $p < 0.001$) and direction (F (2, 48) = 132.916, $p < 0.001$). In particular, healthy controls walked with longer strides when compared to patients, whereas no differences were observed among the sides in the pathological group. All groups reduced their stride lengths during the dual task condition, and the 120° turn showed the lowest value compared with the other conditions.

Additionally, the walking velocity changed with group (F (2, 24) = 2.764, $p < 0.05$), task (F (1, 24) = 21.818, $p < 0.001$) and direction (F (2, 48) = 104.061, $p < 0.001$). In general, controls walked faster than patients, and the latter group showed similar values in both sides. Furthermore, the larger the turning the lower the velocity: the 120° turn showed the lowest velocity value compared to other directions.

The calculated LRI index significantly changed with group (F (2, 24) = 4.672, $p < 0.05$), task (F (1, 24) = 14.712, $p < 0.001$) and direction (F (2, 48) = 70.937, $p < 0.001$). Because of the lack of a significant difference, the data of the PDNA and PDA were pooled together. Our data confirmed that patients walked with a lower LRI compared to controls in all the investigated conditions because of their reduced walking speed, and such a pattern is more pronounced at turning 120° (Figure 4).

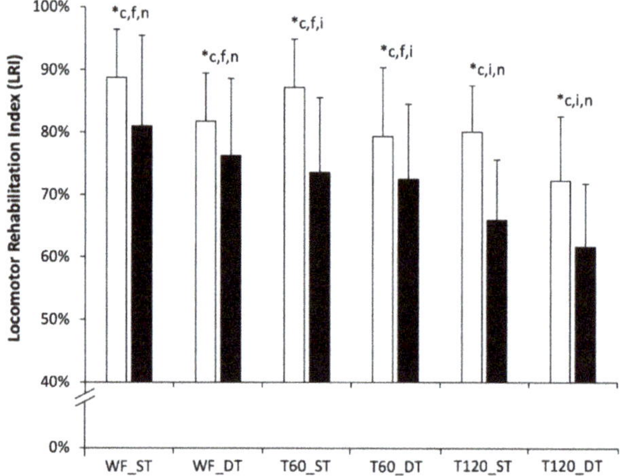


**Figure 4.** Comparison of locomotor rehabilitation index values between the control group (white bars) and the pathological group (black bars), for the forward direction (WF), turning 60° (T60) and turning 120° (T120), in both single (ST) and dual (DT) tasks. Average values ± SD have been reported; data of the PDNA and PDA were pooled together. A significant post-hoc test has been reported: * = $p < 0.05$ by comparing the control group to the pathological one; c = $p < 0.001$: significant difference by comparing the ST to DT; f = $p < 0.001$: significant difference by comparing WF to T60; i = $p < 0.001$: significant difference by comparing T60 to T120; n = $p < 0.001$: significant difference by comparing WF to T120 (same legend as in Tables 2 and 3).

Finally, iso-velocity curves were calculated for each group and condition to understand the complex interaction between stride length and frequency in determining the walk velocity (Figure 5).

During the straight gait (Figure 5, panel a), in single tasking, the patients with PD used the same step frequency compared to controls, but they walked with a shorter stride length. As a consequence, their walking velocities were lower. During the dual task, all curves moved down and to the left: this implied a decrease in the walking speed for controls and patients, and the ratio between the stride length and step frequency was the same.

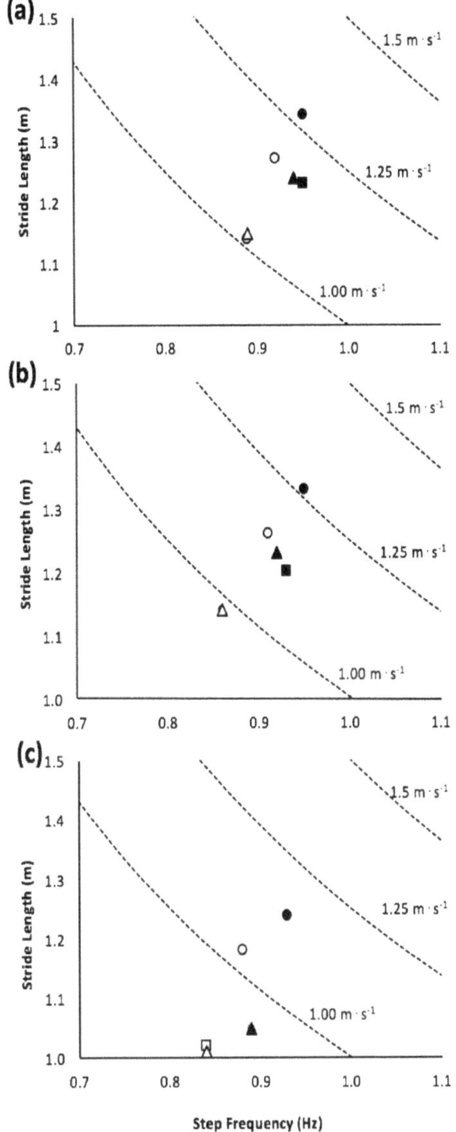

**Figure 5.** Stride length is plotted as a function of the step frequency (i.e., the iso-velocity curves): (**a**) walking forward (WF); (**b**) turning at 60° (T60); (**c**) turning at 120° (T120). The symbols refer to different groups as follows: CRs in the single task •; CRs in the dual task ○; the PDNA in the single task ■; PDNA in the dual task □; PDA in the single task ▲; PDA in the dual task △. For PDNA sometimes the symbols are not seen because they exactly coincide with the data of the other two groups.

This trend occurs similarly in turning (Figure 5, panels b and c), independently of group and task. Importantly, the most significant reduction in the walking velocity occurred while turning 120° and confirms that a larger angle of turning leads to a lower walking velocity.

## 4. Discussion

The present study focused on the effects of dual task and turning on the kinematic parameters in mild to moderate Parkinson's disease. We hypothesized that the simple cognitive and mechanical tasks will display similar effects in both populations, while the combination of these two stimuli will show a higher impact on patients. We found that: *(i)* the temporal walking parameters were affected by the mental task, as well as by the mechanical demand (turning), but no significant differences among populations were observed. On the contrary, the stride length and walking speed were lower in PD patients compared to controls; *(ii)* the turning task had the capability to alter the walking parameters, especially in people with PD, and the major changes in the walking strategy have been observed while turning at a larger angle (120°); *(iii)* the combination of the cognitive and the mechanical task was challenging for patients. Indeed, their stride length and walking velocity showed significant alterations compared to controls in all walking conditions; finally, *(iv)* no significant differences were observed between the not affected/more affected side in all the investigated parameters, suggesting an equal symmetry between the right and left body side.

Taking them together, our results highlighted that a simple mental task alone is not sizeable enough to alter the walking strategy in patients with PD with mild to moderate impairments, whereas the combination of this cognitive task with a change of direction has the capability to modify the walking strategy, especially with a higher turning angle (120°).

### 4.1. Effects on Gait Variables: A Task Comparison

In single tasking, our data confirmed that patients with mild to moderate PD walked with a reduced gait velocity and a shorter step length compared to controls. Therefore, patients showed a lower locomotor index rehabilitation if compared to controls with no evident differences between the not affected and more affected side. This finding suggests that the mild to moderate pathology compromises, in a similar way, the gait kinematics of the inferior limbs (e.g., it did not compromise the symmetry among body sides). Since, postural control and gait are linked to cognitive function both in healthy and pathological subjects [18,22,42], it is possible to assume that PD patients reduced their stride length in order to increase the time spent on the ground, increasing the walking stability.

As we hypothesized, a simple dual task condition (repeating the days of the week backwards) affected both groups similarly: all participants increased the ground contact (and the cycle time) and decreased their cadence and frequency [12,19]. This finding endorses that subjects tried to maintain their postural stability by spending more time on the ground, as this would prevent the risk of falling [18]. The lack of a difference among groups confirms that, by adding a cognitive load, a low disease severity could not play a major role in determining motor impairments [23]. In addition, the high focus on the additional task means a larger proportion of the attentional capacity is at the expense of walking performance: people walked even more slowly with much shorter steps [7,12,15,22,43–46]. This "compensatory strategy" could be useful in achieving a greater control of gait and balance disruption [20], and in counterbalancing the fluctuations of the center of mass.

Finally, the similar ratio between the stride length and step frequency means that subjects did not change their stride pattern. This latter outcome suggests that the simple cognitive task was probably not too demanding for all participants. More complex cognitive tasks (i.e., concurrent loads, mental tracking) would probably point out the gap among groups [19,20].

### 4.2. Effects on Gait Variables: A Direction Comparison

In the present study, we investigated two turning angles that are very common during daily activity [38]: the first (60°) represents a simple turn and the other (120°) a more difficult turn.

Our spatiotemporal data are in line with the previous literature [8,10,18,21]. With respect to directions, whereas healthy participants were impacted by the 120° turn, PD patients were also affected by the 60° turn, but only for the walking velocity. Additionally, in this case, no appreciable differences between the not affected and more affected side were observed.

Generally, the 120° turning is characterized by an increased support time and a reduced number of strides (i.e., cadence) and frequency, as well as a decreased speed and length. This angle requires additional attentional resources: in fact, it relies more heavily on proprioception (i.e., directed by the basal ganglia function) than both forward walking and the 60° turn [47]. By considering the mechanical approach, the impact of a larger turning could be explained by the kinetics of the movement. Indeed, to move into a larger angle of turn, the subjects decreased their speed more (just for a greater angle of turn) and, probably, used their turn foot as "pivot foot". When the turn angle increased, the ankle plantar flexion moment (and its peak) increased and the external rotators of the lower extremity played a much greater role than in straight walking [48–50]. These alterations of the plantar flexors have been accomplished with a reduced ankle power generated in the pre-swing phase (push-off power). Therefore, the larger the angle of turn the higher the mechanical demand. As a result, PD patients tended to reduce the step length and to increase the contact time primarily to improve the stability of the body. They showed an initial alteration of the walking pattern at a 60° turn, but only the walking velocity was affected. On the other hand, a 120° turn showed evident differences when compared to the straight direction in terms of both the temporal and spatial parameters; this finding supports the idea of a more challenging stimulus.

Therefore, whereas the 60° turn could represent a not suitable training stimulus, the 120° may be a good challenge for people with PD. As expected, all gait alterations are exacerbated during walking with a combination of the two stimuli, especially in PD patients.

*4.3. Effects on Gait Variables: Cognitive Task and Mechanical Task*

The combined effects of dual (cognitive) and mechanical (turn) tasks represent an important training stimulus in people with PD. Indeed, our data shows that the concomitant presence of a simple mental and turning (60°) task produced only a marginal effect on the main kinematics. In particular, the timing parameters showed no significant differences compared with the straight line, and the differences between the single and dual task are comparable to the ones obtained for healthy subjects. Furthermore, even the step length and the step frequency showed a similar trend.

On the contrary, the matching between the same mental task and a more complex mechanical demand (120° turn) played a greater "destructive" impact especially in patients with PD. Indeed, their temporal and spatial variables showed a much more marked gap than the controls' ones.

Our results were based on a relatively small heterogeneous sample of disabled patients, who lived independently in the community. Therefore, further researches are necessary to extend these findings to a larger sample (i.e., patients with more severe gait deficits or episodes of freezing), or to other conditions (i.e., OFF phase performance).

## 5. Conclusions

Our data showed that a mechanical task (i.e., turning) has the potential to modify gait strategy in people with Parkinson's disease, without changes in symmetry of the lower limbs. Of greatest interest, the concomitant presence of a mechanical task and a simple cognitive task did not produce a further impairment of this gait strategy. Therefore, using the investigated combined condition (turning and repeating the days of the week backwards) could represent a significant training stimulus in such patients. Indeed, the improvement of the mental and physical characteristics is very important in improving the functionality of patients at the early stages of their pathology.

**Author Contributions:** Conceptualization, F.N., F.B., A.M., M.B.; methodology, F.N., A.M.; software, F.N., E.B.; data curation, F.N., E.B., A.M.; writing—original draft preparation, F.N., A.M., M.B.; writing—review and editing, F.N., A.M., M.B. All authors have read and agreed to the published version of the manuscript.

**Funding:** This research received no external funding.

**Acknowledgments:** We would like to thank Alessandro Corsi, Gianluca Fedel and Davide Nisi for their help in data collection, and the subjects for participating in the study.

**Conflicts of Interest:** The authors acknowledge that there are no conflicts of interest pertaining to this manuscript.

## References


1. De Lau, L.M.; Breteler, M.M. Epidemiology of Parkinson's disease. *Lancet Neurol.* **2016**, *5*, 525–535. [CrossRef]
2. Kelly, V.E.; Eusterbrock, A.J.; Shumway-Cook, A. A review of dual-task walking deficits in people with Parkinson's disease: Motor and cognitive contributions, mechanisms, and clinical implications. *Parkinsonism Disord.* **2012**, *2012*, 918719. [CrossRef] [PubMed]
3. Cummings, J.L. Intellectual impairment in Parkinson's disease: Clinical, pathological, and biochemical correlates. *J. Ger. Psychiatry Neurol.* **1988**, *1*, 24–36. [CrossRef] [PubMed]
4. Creaby, M.W.; Cole, M.H. Gait characteristics and falls in Parkinson's disease: A systematic review and meta-analysis. *Parkinsonism Related Disord.* **2018**, *57*, 1–8. [CrossRef]
5. Kelly, V.E.; Eusterbrock, A.J.; Shumway-Cook, A. The effects of instructions on dual-task walking and cognitive task performance in people with Parkinson's disease. *Parkinsonism Dis.* **2012**. [CrossRef]
6. Nocera, J.R.; Roemmich, R.; Elrod, J.; Atlmann, L.J.P.; Hass, C.J. Effects of cognitive task on gait initiation in Parkinson disease: Evidence of motor prioritization? *J. Rehabil. Res. Develop.* **2013**, *50*, 699–708. [CrossRef]
7. Rochester, L.; Galna, B.; Lord, S.; Burn, D. The nature of dual-task interference during gait in incident Parkinson's disease. *Neuroscience* **2014**, *265*, 83–94. [CrossRef]
8. Morris, M.E.; Iansek, R.; Matyas, T.A.; Summers, J.J. Stride length regulation in Parkinson's disease: Normalization strategies and underlying mechanisms. *Brain* **1996**, *119*, 551–568. [CrossRef]
9. Morris, M.E.; Huxham, F.; McGinley, J.; Dodd, K.; Iansek, R. The biomechanics and motor control of gait in Parkinson disease. *Clin. Biomech.* **2001**, *16*, 459–470. [CrossRef]
10. Canning, C.G. The effect of directing attention during walking under dual task conditions in Parkinson's disease. *Parkinsonism Relat. Disord.* **2005**, *11*, 95–99. [CrossRef]
11. Plotnik, M.; Dagan, Y.; Gurevich, T.; Giladi, N.; Hausdorff, J.M. Effects of cognitive function on gait and dual tasking abilities in patients with Parkinson's disease suffering from motor response fluctuations. *Exp. Brain Res.* **2011**, *208*, 169–179. [CrossRef] [PubMed]
12. O'Shea, S.; Morris, M.E.; Iansek, R. Dual task interference during gait in people with Parkinson disease: Effects of motor versus cognitive secondary tasks. *Phys. Ther.* **2002**, *82*, 888–897. [CrossRef] [PubMed]
13. Rochester, L.; Hetherington, V.; Jones, D.; Nieuwboer, A.; Willems, A.; Kwakkel, G.; van Wegen, E. Attending to the task: Interference effects of functional tasks on walking in Parkinson's disease and the roles of cognition, depression, fatigue and balance. *Arch. Phys. Med. Rehab.* **2004**, *85*, 1578–1585. [CrossRef] [PubMed]
14. Brauer, S.G.; Morris, M.E. Can people with Parkinson's disease improve dual tasking when walking? *Gait Posture* **2010**, *31*, 229–233. [CrossRef] [PubMed]
15. Brown, L.A.; de Bruin, N.; Doan, J.B.; Suchowersky, O.; Hu, B. Novel challenges to gait in Parkinson's disease: The effect of concurrent music in single and dual task contexts. *Arch. Phys. Med. Rehab.* **2009**, *90*, 1578–1583. [CrossRef] [PubMed]
16. Salazar, R.D.; Ren, X.; Ellis, T.; Toraif, N.; Barthelemy, O.J.; Neargarder, S.; Cronin-Golomb, A. Dual tasking in Parkinson's disease: Cognitive consequences while walking. *Neuropsychology* **2016**, *31*, 613–623. [CrossRef]
17. Campbell, C.M.; Rowse, J.L.; Ciol, M.A.; Shumway-Cook, A. The effect of cognitive demand on timed up and Go performance in older adults with and without Parkinson disease. *Neurol. Rep.* **2003**, *27*, 2–7. [CrossRef]
18. Camicioli, R.; Oken, B.S.; Sexton, G.; Kaye, J.A.; Nutt, J.G. Verbal fluency task affects gait in Parkinson's disease with motor freezing. *J. Geriatr. Psych. Neurol.* **1998**, *11*, 181–185. [CrossRef]
19. Galletly, R.; Brauer, S.G. Does the type of concurrent task affect preferred and cued gait in people with Parkinson's disease. *Austr. J. Physiother.* **2005**, *51*, 175–180. [CrossRef]
20. Lapointe, L.L.; Stierwalt, J.A.G.; Maitland, C.G. Talking with walking: Cognitive loading and injurious falls in Parkinson's disease. *Intern. J. Speech-Lang. Pathol.* **2010**, *12*, 445–459. [CrossRef]
21. Stegemoller, E.L.; Wilson, J.P.; Hazamy, A.; Shelly, M.C.; Okun, M.S.; Altmann, L.P.; Hass, C.J. Associations between cognitive and gait performance during single- and dual-task walking in people with Parkinson disease. *Phys. Ther.* **2014**, *94*, 757–766. [CrossRef] [PubMed]

22. Yogev, G.; Giladi, N.; Peretz, C.; Springer, S.; Simon, E.S.; Hausdorff, J.M. Dual tasking, gait rhythmicity, and Parkinson's disease: Which aspects of gait are attention demanding? *Eur. J. Neurosci.* **2005**, *22*, 1248–1256. [CrossRef] [PubMed]

23. Wild, L.B.; de Lima, D.B.; Balardin, J.B.; Rizzi, L.; Giacobbo, B.L.; Oliveira, H.B.; de Lima, A.; Peyre-Tartaruga, L.A.; Rieder, C.R.; Bromberg, E. Characterization of cognitive and motor performance during dual-tasking in healthy older adults and patients with Parkinson's disease. *J. Neurol.* **2013**, *260*, 580–589. [CrossRef] [PubMed]

24. Hong, M.; Perlmutter, J.S.; Earhart, G.M. A kinematic and electromyographic analysis of turning in people with Parkinson disease. *Neurorehabil. Neural Repair* **2009**, *23*, 166–176. [CrossRef] [PubMed]

25. Stack, E.; Jupp, K.; Ashburn, A. Developing methods to evaluate how people with Parkinson's disease turn 180°: An activity frequently associated with falls. *Disab. Rehabil.* **2004**, *26*, 478–484. [CrossRef]

26. Stack, E.L.; Ashburn, A.M.; Jupp, K.E. Strategies used by people with Parkinson's disease who report difficulty turning. *Parkinson. Relat. Disord.* **2006**, *12*, 87–92. [CrossRef]

27. Adamson, M.B.; Gilmore, G.; Stratton, T.W.; Baktash, N.; Jog, M.S. Medication status and dual tasking on turning strategies in Parkinson disease. *J. Neurol. Sci.* **2019**, *396*, 206–212. [CrossRef]

28. Chou, P.; Lee, S. Turning deficits in people with Parkinson's disease. *Tzu Chi Med. J.* **2013**, *25*, 200–202. [CrossRef]

29. Vaugoyeau, M.; Viallet, F.; Mesure, S.; Massion, J. Coordination of axial rotation and step execution: Deficits in Parkinson's disease. *Gait Posture* **2003**, *18*, 150–157. [CrossRef]

30. Crenna, P.; Carpinella, I.; Rabuffetti, M.; Calabrese, E.; Mazzoleni, P.; Nemni, R.; Ferrarin, M. The association between impaired turning and normal straight walking in Parkinson's disease. *Gait Posture* **2007**, *26*, 172–178. [CrossRef]

31. Willems, A.M.; Nieuwboer, A.; Chavret, F.; Desloovere, K.; Dom, R.; Rochester, L.; Kwakkel, G.; van Wegen, E.; Jones, D. Turning in Parkinson's disease patients and controls: The effect of auditory cues. *Mov. Disord.* **2007**, *22*, 1871–1878. [CrossRef] [PubMed]

32. Mellone, S.; Mancini, M.; King, L.A.; Horak, F.B.; Chiari, L. The quality of turning in Parkinson's disease: A compensatory strategy to prevent postural instability? *J. Neuroeng Rehabil.* **2016**, *13*, 39. [CrossRef] [PubMed]

33. Bertoli, M.; Della Croce, U.; Cereatti, A.; Mancini, M. Objective measures to investigate turning impairments and freezing of gait in people with Parkinson's disease. *Gait Posture* **2019**, *74*, 187–193. [CrossRef] [PubMed]

34. Encarna Micó-Amigo, M.; Kingma, I.; Heinzel, S.; Nussbaum, S.; Heger, T.; Lummel, R.C.V.; Dieën, J.H.V. Dual vs. Single Tasking during Circular Walking: What Better Reflects Progression in Parkinson's Disease? *Front. Neurol.* **2019**. [CrossRef]

35. Spildooren, J.; Vercruysse, S.; Desloovere, K. Freezing of gait in Parkinson's disease: The impact of dual-tasking and turning. *Mov. Disord.* **2010**, *25*, 2563–2570. [CrossRef] [PubMed]

36. Bayot, M.; Dujardin, K.; Tard, C.; Defebvre, L.; Bonnet, C.T.; Allart, E.; Delval, A. The interaction between cognition and motor control: A theoretical framework for dual-task interference effects on posture, gait initiation, gait and turning. *Clin. Neurophys.* **2018**, *48*, 361–375. [CrossRef]

37. Hackney, M.E.; Earhart, G.M. The effects of a secondary task on forward and backward walking in Parkinson disease. *Neurorehabil. Neural Repair* **2010**, *24*, 97–106. [CrossRef]

38. Patla, A.E.; Prentice, S.D.; Robinson, C.; Neufeld, J. Visual control of locomotion: Strategies for changing direction and for going over obstacles. *J. Exp. Psychol.* **1991**, *17*, 603–634. [CrossRef]

39. Hughes, A.J.; Daniel, S.E.; Kilford, L.; Lees, A.J. Accuracy of clinical diagnosis of idiopathic Parkinson's disease: A clinico-pathological study of 100 cases. *J. Neurol. Neurosurg. Psychiatry* **1992**, *55*, 181–184. [CrossRef]

40. Schenkman, M.L.; Clark, K.; Xie, T.; Kuchibhatla, M.; Shinberg, M.; Ray, L. Spinal movement and performance of a standing reach task in participants with and without Parkinson disease. *Phys. Ther.* **2001**, *81*, 1400–1411. [CrossRef]

41. Movement Disorder Society Task Force on Rating Scales for Parkinson's Disease. The Unified Parkinson's Disease Rating Scale (UPDRS): Status and recommendations. *Mov. Disord.* **2003**, *18*, 738–750. [CrossRef] [PubMed]

42. Peyre-Tartaruga, L.A.; Monteiro, E.P. A new integrative approach to evaluate pathological gait: Locomotor rehabilitation index. *Clin. Transl. Degener. Dis.* **2016**, *1*, 86–90. [CrossRef]

43. Woollacott, M.; Shumway-Cook, A. Attention and the control of posture and gait: A review of an emerging area of research. *Gait Posture* **2002**, *16*, 1–14. [CrossRef]

44. Plotnik, M.; Giladi, N.; Hausdorff, J.M. Bilateral coordination of gait and Parkinson's disease: The effects of dual tasking. *J. Neurol. Neurosurg. Psych.* **2009**, *80*, 347–350. [CrossRef]

45. Hausdorff, J.M.; Balash, J.; Giladi, N. Effects of cognitive challenge on gait variability in patients with Parkinson's disease. *J. Geriatr. Psych. Neurol.* **2003**, *16*, 53–58. [CrossRef] [PubMed]

46. De Melo Roiz, R.; Avezedo Cacho, E.W.; Macedo Pazinatto, M.; Guimaraes Reis, J.; Cliquet, A.; Barasnevicius-Quagliato, E.M.A. Gait analysis comparing Parkinson's disease with healthy elderly subjects. *Arq. Neuropsiquiatr.* **2010**, *68*, 81–86. [CrossRef]

47. Sofuwa, O.; Nieuwboer, A.; Desloovere, K.; Willems, A.M.; Chavret, F.; Jonkers, I. Quantitative gait analysis in Parkinson's disease: Comparison with a healthy control group. *Arch. Phys. Med. Rehabil.* **2005**, *86*, 1007–1013. [CrossRef]

48. Taylor, M.J.D.; Dadbnichki, P.; Strike, S.C. A three-dimensional biomechanical comparison between turning strategies during the stance phase of walking. *Hum. Mov. Sci.* **2005**, *24*, 558–573. [CrossRef]

49. Segal, A.; Orendurff, M.; Flick, K.; Berge, J.; Klute, G. Ankle biomechanics during a spin turn. In Proceedings of the 28th Annual Meeting of the American Society of Biomechanics, Boulder, CO, USA, 15 July 2004; p. 168.

50. Xu, D.; Chow, J.W.; Wang, Y.T. Effects of turn angle and pivot foot on lower extremity kinetics during walk and turn actions. *J Appl. Biomech.* **2006**, *22*, 74–79. [CrossRef]

*Article*

# Race Walking Ground Reaction Forces at Increasing Speeds: A Comparison with Walking and Running

**Gaspare Pavei** [1,*], **Dario Cazzola** [2], **Antonio La Torre** [3] and **Alberto E. Minetti** [1]

[1]  Laboratory of Physiomechanics of Locomotion, Department of Pathophysiology and Transplantation, University of Milan, Via Mangiagalli 32, I-20133 Milan, Italy
[2]  Department for Health, University of Bath, Bath BA2 7AY, UK
[3]  Department of Biomedical Sciences for Health, University of Milan, Milano, Via G. Colombo 71, I-20133 Milan, Italy
*  Correspondence: gaspare.pavei@unimi.it

Received: 10 June 2019; Accepted: 1 July 2019; Published: 3 July 2019

**Abstract:** Race walking has been theoretically described as a walking gait in which no flight time is allowed and high travelling speed, comparable to running (3.6–4.2 m s$^{-1}$), is achieved. The aim of this study was to mechanically understand such a "hybrid gait" by analysing the ground reaction forces (GRFs) generated in a wide range of race walking speeds, while comparing them to running and walking. Fifteen athletes race-walked on an instrumented walkway (4 m) and three-dimensional GRFs were recorded at 1000 Hz. Subjects were asked to performed three self-selected speeds corresponding to a low, medium and high speed. Peak forces increased with speeds and medio-lateral and braking peaks were higher than in walking and running, whereas the vertical peaks were higher than walking but lower than running. Vertical GRF traces showed two characteristic patterns: one resembling the "M-shape" of walking and the second characterised by a first peak and a subsequent plateau. These different patterns were not related to the athletes' performance level. The analysis of the body centre of mass trajectory, which reaches its vertical minimum at mid-stance, showed that race walking should be considered a bouncing gait regardless of the presence or absence of a flight phase.

**Keywords:** body centre of mass; human gait; race walking; force plate

---

## 1. Introduction

Ground reaction forces (GRFs) have often been used in biomechanical studies to describe human locomotion [1,2] because they show that the forces exerted by the foot on the ground are a key determinant of the final gait kinematics. Ground reaction force analysis is nowadays used also for the detection of pathological gaits [3,4], gait asymmetry [5,6], injury prevention [7] and in the estimation of muscles force [8]. Moreover, the analysis of GRF peaks and the timing of peaks occurrence can explain how velocity is generated and increased, which could be important in sport activity such as running (e.g., [9]).

Ground reaction forces double integration is also used to compute body centre of mass (BCoM) trajectory and describe locomotion mechanics [10]. In race walking, the BCoM trajectory can be correctly computed only by using a forward dynamics approach, whereas inverse dynamics computation has been shown to be biased [11]. Thus, the measurement and analysis of GRFs at increasing speed is key to investigating race walking mechanics, even more than in walking and running. Each animal or human gait has its own "locomotor signature", ultimately represented by the trajectory of the body centre of mass (BCoM), with its asymmetry and related energies [12], and race-walking trajectory has never been analysed in such a fashion due to the lack of consistent GRF datasets. Starting from the ground reaction forces recorded during a stride, it is also possible to represent the "locomotor signature" by showing the Lissajous contour [12] also for race walking gait.

Race walking is an Olympic discipline where athletes are required to complete the distance in the shortest time according to two constrains: no flight time can occur between steps and the knee has to be locked in extension from touch down to mid-stance. This rule induces race walking to manage a different kinematics compared with walking and running that causes, also, somewhat different ground reaction forces patterns [13–15]. However, race walking dynamics have been less studied when compared with walking and running [16], and often investigations have been focused on one speed only, missing potentially relevant information about velocity generation.

The aim of this study was to analyse the ground reaction forces and BCoM trajectory during race walking on a wide range of speeds and to compare the three components of GRFs (i.e., forward, lateral, vertical) with walking and running.

## 2. Materials and Methods

Fifteen male athletes (mean ± SD, 23.0 ± 5.5 years old, 1.78 ± 0.05 m height, 64.7 ± 5.2 kg body mass and with a 10,000 m personal best of 44:26 ± 3:34 min:s) participated in this study. All subjects gave their informed consent for inclusion before they participated in the study. The study was conducted in accordance with the Declaration of Helsinki, and the protocol was approved by the Ethics Committee of the University of Milan.

The "dynamometric corridor" was composed of five 3D strain gauge platforms (Bertec, USA) in order to obtain a 4 m long and 0.4 m wide footpath placed in the middle of a 40 m walkway where athletes could race walk at a constant speed. Athletes were asked to perform three trials at each self-chosen low, medium and high speed, hence, each subject completed 9 trials.

Ground reaction forces were recorded at 1000 Hz and normalised to athlete's body weight (BW) and as a percentage of stance time. Stance phase was defined using a 10 N threshold on vertical ($F_Z$) force. The inversion between braking and propulsive on antero-posterior ($F_X$) force was set when the force from negative (braking) became positive (propulsive). Speeds were clustered when within 3% of the target speed (2.78, 3.06, 3.33, 3.61, 3.89, 4.17 m s$^{-1}$), similar to the cluster used in Nillson and Thorstensson [17]. Walking and running GRF values were also taken from the Nillson and Thorstensson [17] study for comparison purposes.

The BCoM position was computed by double integration of the 3D acceleration, obtained by the force signal, according to Cavagna [10]; the integration constants were calculated as described in the Appendix of Saibene and Minetti [18]. The obtained BCoM trajectory was transformed in local coordinates (as the sampling occurred over an instrumented treadmill). The resulting 3D contour included several consequent strides, each of which was forced to become a closed loop and centred on (0, 0, 0) by subtracting average 3D coordinates to allow a description based on a Fourier Series with 6 harmonics [12]. Walking and running BCoM data were extracted from our cumulated database on subjects matched for anthropometry and age ($n = 10$).

Statistical differences across speeds and multiple peaks were tested by a two-way ANOVA using a Bonferroni post-hoc test, whereas differences between the two main gait patterns (see Results) were tested using a $t$-test and the significant level was set at $p < 0.05$ (SPSS 19, IBM).

## 3. Results

Mean curves of vertical (Fz), antero-posterior (Fx) and medio-lateral (Fy) ground reaction forces during race walking stance phase are presented in Figure 1. Subjects were clustered in two different groups according to different Fz curves: (i) M shape (Figure 1) that was similar to walking, displaying two peaks and a valley and (ii) N shape (Figure 1) that showed a huge first peak followed by a plateau. The Fx curve showed a first negative braking peak and a smaller propulsive peak near the end of stance phase, without substantial differences on the Fx curve among the two groups. Braking and propulsive impulse (i.e., the area of the two phases delimited by the abscissa at 0 value) were always very similar denoting a substantially constant inter-strides speed. The Fy curves were averaged between right and left stance since they were specular with no significant differences. After a first medial small peak, the force was lateral, medial at mid stance and lateral again at two-thirds of the stance. The two groups reached the first lateral peak with two different shapes, more consistently for the M shape group.

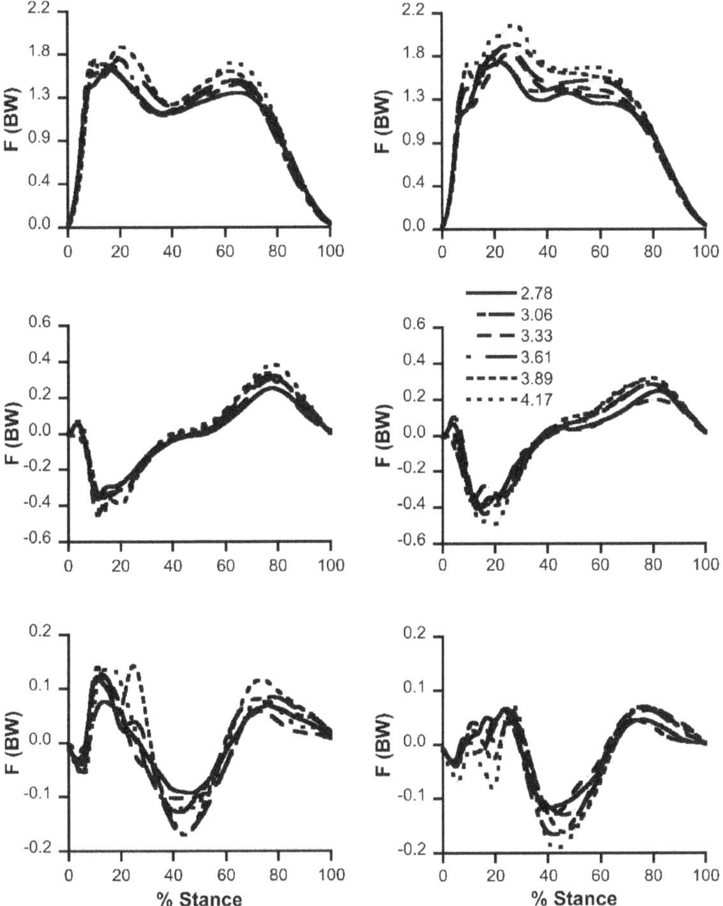

**Figure 1.** Ground reaction force traces (as a fraction of body weight, BW) in the three axes (vertical, antero-posterior and medio-lateral axis from top to bottom) at increasing speed (m s$^{-1}$) are shown. Right and left columns represent the M shape and N shape vertical force patterns, respectively.

In Figure 2, vertical (Fz) peaks and valleys in race walking, walking and running [17] are presented. When analysing the M shape group, the first peak was higher than walking, whereas the second was comparable, and the valley instead was much higher in race walking than in walking without falling under the BW value. In the M shape group, the first Fz peak was always significantly higher ($p < 0.001$) than the second. The N shape Fz peak was slightly higher than the M shape, with significant difference only at 2.78 m s$^{-1}$ ($p < 0.01$) and 3.61 m s$^{-1}$ ($p < 0.05$). Running showed higher peaks than the other gaits. All peaks increased linearly with speed, with significant differences between 4.17 m s$^{-1}$ and the other speeds in the N shape and the second M shape peak ($p < 0.01$). The first M peak increased with speed but with a less significant trend, whilst the valley was speed independent.

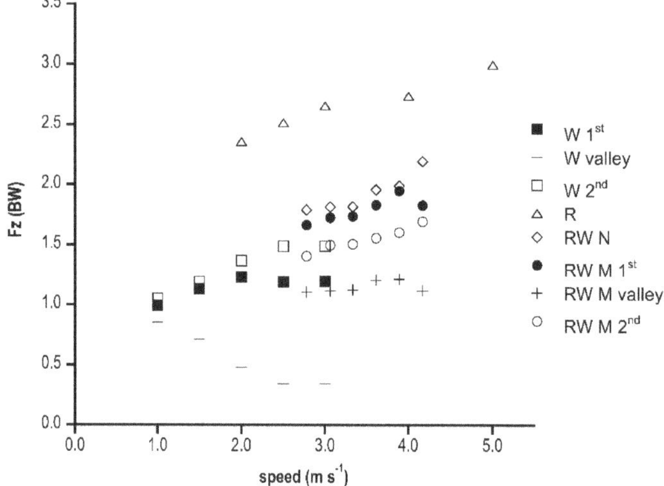

**Figure 2.** Ground reaction force vertical peaks (as a fraction of BW) in the three gaits (W, walking; R, running; RW, race walking) at increasing speed (m s$^{-1}$) are shown. M and N refer to race walking M shape and N shape groups, respectively.

Figure 3 shows the antero-posterior (Fx) peaks in the three gaits. Braking and propulsive peaks increased linearly with speed and with the same values in walking and running, whereas race walking reported higher braking than propulsive peaks ($p < 0.001$) in both groups at each speed. The braking peak was significantly lower ($p < 0.05$) in the M shape than the N shape except for 3.33 m s$^{-1}$ where the M shape peak was higher and at 3.89 m s$^{-1}$ that was not different.

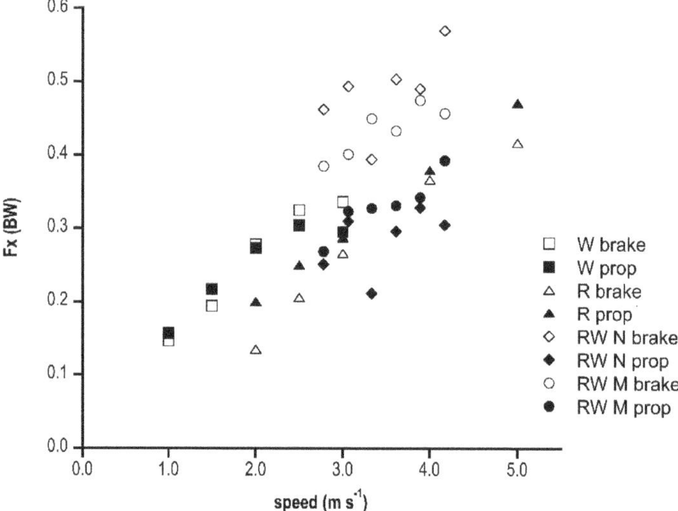

**Figure 3.** Antero-posterior peaks (as a fraction of BW) of ground reaction force in the three gaits (W, walking; R, running; RW, race walking) at increasing speed (m s$^{-1}$) are shown. "Brake" is braking peak; "prop" represents propulsive peak. M and N refer to race walking M shape and N shape groups, respectively.

In Figure 4A, the difference between Fy medial and lateral peak are shown: walking and running Fy increased in a similar fashion linearly with speed, and race walking always showed higher values (significant differences among groups only at 3.89 m s$^{-1}$ ($p < 0.001$)); in Figure 4B, the amplitude of medial and lateral peaks in race walking were almost speed independent. Also, across several speeds, the first lateral peak was significantly higher than the second one. In the N shape group, the medial peak was also greater than the lateral, whereas in the M shape group, the lateral and medial force peaks were comparable. The N shape group often showed significantly greater peak values than the M shape one.

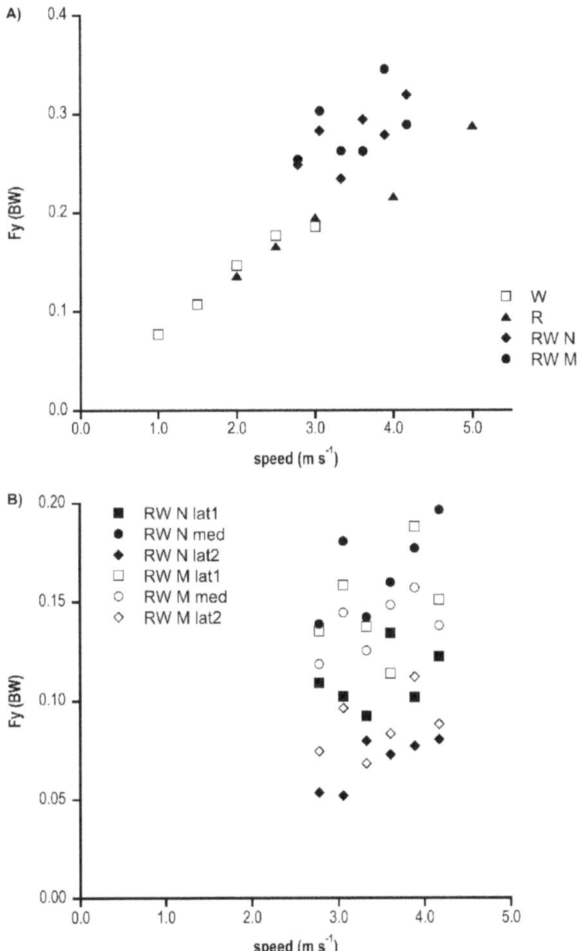



**Figure 4.** (**A**) Ground reaction medio-lateral "delta" force (as a fraction of BW), expressed as the peak-to-peak force difference, for the three gaits (W, walking; R, running; RW, race walking) at increasing speed (m s$^{-1}$) are shown. (**B**) Peak medial (med) and lateral (lat) force (as a fraction of BW) during race walking at increasing speed (m s$^{-1}$). M and N refer to race walking M shape and N shape groups, respectively.

Figure 5A (M shape group) and b (N shape group) shows the timing of the peaks, valley and inversion of the vertical, antero-posterior and medio-lateral ground reaction forces in relation to the normalised stance phase for each speed tested. Most of the variables did not show speed dependency, since their relative timing across speeds did not change; however, in the N shape group, the propulsive peak at 4.17 m s$^{-1}$ occurred significantly earlier ($p < 0.05$) than in other speeds. In the M shape group, the peak brake and peak lateral forces showed some timing variations, ($p < 0.05$). It was interesting to note that some peaks occurred together: Fx brake, Fy lateral and Fz; medial Fy and Fx inversion; and Fx propulsive and Fy lateral, without differences among groups. The M shape group showed a later inversion of antero-posterior force at speed $< 3.89$ m s$^{-1}$ ($p < 0.01$), an earlier propulsive peak at some speeds ($p < 0.01$), an earlier lateral peak at speed $< 3.61$ m s$^{-1}$ ($p < 0.05$) and an earlier medial peak at high speed ($p < 0.05$) than the N shape group.

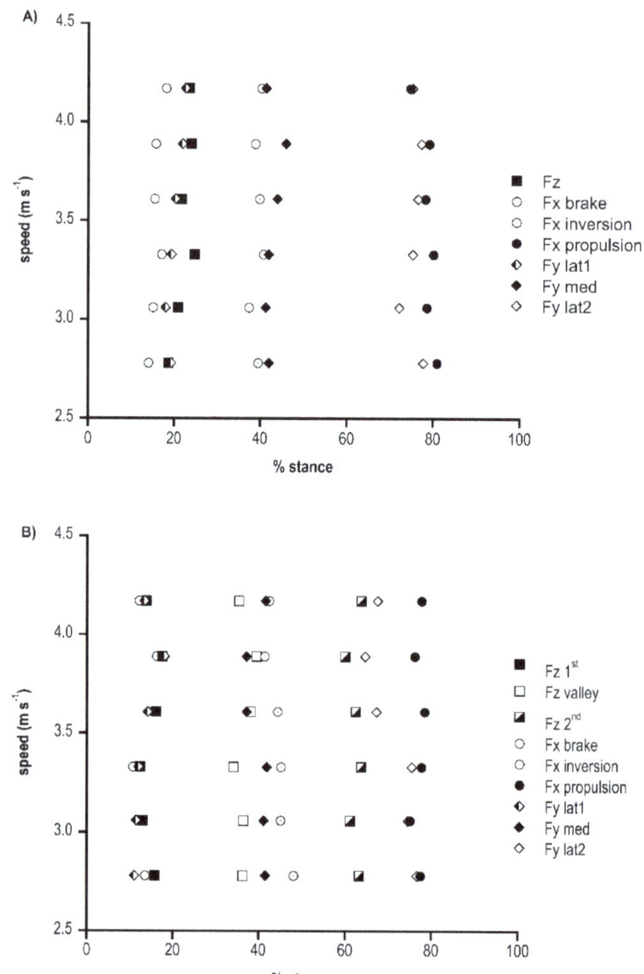

**Figure 5.** Timing of race walking force peaks (%stance) at the different speeds (m s$^{-1}$). (**A**) N shape group. (**B**) M shape group.

By comparing race walking peaks' timing with walking and running [17], braking and vertical peaks were anticipated in race walking, whereas propulsive peak timing was similar to walking and occurred later than running.

The 3D BCoM trajectory is presented in Figure 6 in comparison with running at the same speed and walking (1.94 m s$^{-1}$). The race walking BCoM volume was smaller than in running and walking, with a narrower displacement in the medio-lateral direction and smaller vertical excursion. Walking and running showed a lower minimum of the BCoM vertical trajectory compared with race walking, whereas in the upper part, walking and race walking showed the same maximum, which was lower than in running. In race walking, BCoM was in the lowest part of the trajectory during stance, as in running, without showing the arc of circle characteristic of walking during the stance phase. The set of equations, based on the Fourier Series (truncated at the 6th harmonic), needed to describe the BCoM 3D trajectory of race walking (for example at 3.61 m s$^{-1}$) is:

$$\begin{bmatrix} x \\ y \\ z \end{bmatrix} = \begin{bmatrix} 5.017sin(2t - 0.609) + 0.296sin(4t - 2.219) + 0.191sin(6t + 1.380) \\ 5.080sin(t + 0.000) + 0.491sin(3t + 2.362) + 0.148sin(5t - 1.778) \\ 14.421sin(2t + 1.250) + 2.069sin(4t + 0.578) + 0.371sin(6t - 0.906) \end{bmatrix}$$
$$+1.865sin(t + 0.101) + 0.130sin(3t - 2.211) + 0.054sin(5t - 2.107)$$
$$+0.203sin(2t - 0.610) + 0.052sin(4t - 1.232) + 0.025sin(6t - 2.281) \Big], t = 0 \dots 2\pi$$
$$+2.957sin(t - 2.110) + 0.554sin(3t + 2.859) + 0.113sin(5t + 2.357)$$

where x, y, and z are the antero-posterior, medio-lateral and vertical axis, respectively. This kind of equation was used to represent the walking, running and race walking BCoM trajectory in Figure 6.

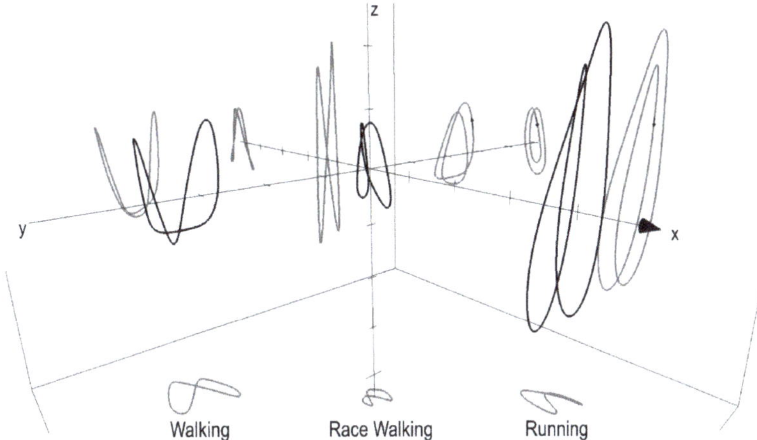


**Figure 6.** 3D representation using the Fourier Series of the Body Center of Mass (BCoM) trajectory of walking (1.94 m s$^{-1}$), race walking and running (2.78 m s$^{-1}$). The mean contours (in black) have been shifted in the antero-posterior and medio-lateral axes (x and y, respectively) to better appreciate the 2D projection (in grey) on each plane. Black arrows on the contour projections in the XZ plane represent the BCoM dynamic movement along the trajectory, which is counter-clockwise for walking and clockwise for race walking and running. Progression direction is shown by the black arrows on the antero-posterior (x) axis, and the axes thickness is 0.02 m.

## 4. Discussion

This paper described and analysed the race walking ground reaction forces patterns in the three planes of motion at increasing speeds. The GRF peak values were comparable with the paired speeds of previous studies in literature [13–15,19], with a few discrepancies which are discussed below.

### 4.1. Speed Adaptation

The speed increased vertical peaks almost linearly and the second peak in the M shape race walking group was more affected than the first one (Figure 2). On the contrary, the valley results were speed independent (Figure 2). The peak propulsive and braking forces also increased with speed and the braking peak was always greater than the propulsive peak, whereas the timing of the inversion was not affected by speed (Figure 3). The braking peak is expected to be passive in race walking as explained by the "locked-knee" rule, but since the braking and propulsive impulses were the same (as they should for the constant speed), such a peak force asymmetry seems to suggest a strategy to minimize peak muscle involvement in the propulsive phase. Medio-lateral forces, both in absolute terms or expressed as peaks difference, were similar at increasing speed (Figure 4), and this suggests that the leg and trunk muscles do not need to increase their activity when speed increases, as the body centre of mass trajectory was not further laterally deviated. Such speculation is supported from a gait optimisation perspective, as a lateral deviation with respect to the progression direction has been shown as an avoidable and functionally ineffective feature in all animal gaits (apart from penguins [20]). Timing of peaks, when expressed as stance percentage, was speed independent except for few variations (Figure 5). This is a typical gait stereotype, which could be beneficial from a motor learning perspective [21], since, in this way, athletes do not have to change their pattern to gain speed but only perform it faster. In fact, when increasing speed, the contact time is reduced [22], and the gait events are anticipated in absolute timing. Since the majority of the peaks increased with speed, as occurred in walking and running [17], comparison within the same gait or with other gaits should be performed only at matched speed.

*4.2. Race Walking versus Walking and Running*

When comparing GRFs across race walking, walking and running, some differences were found. Vertical forces were higher in running, whereas race walking peaks seemed to increase with speed following the same trend of walking values at higher speeds (Figure 2). However, while in walking, the valley values dropped under body weight and decreased with speed; in race walking, the valley values were slightly higher than body weight and speed independent (Figure 2). In walking, the trough drops under body weight due to the "centrifugal reaction force" caused by the arc of circumference quickly travelled by BCoM during the stance phase [23,24]. In race walking, the BCoM during stance does not move along a circle with leg length as radius [11], but lowers down as in running. This could explain why the trough in vertical force signal does not drop under body weight; however, this hypothesis does not explain why an M shape pattern is exhibited, as in walking, whilst the BCoM pathway is similar to running. Also, both vertical peaks occurred earlier than walking with a timing similar to running (for the first). When considering the vertical forces, running is the most stressful gait due to the highest peaks. Race walking propulsive peaks were comparable to running, differently, the braking peaks were higher than running (Figure 3). The inversion between braking and propulsive in race walking occurred earlier than in running, and a shorter braking time involves a higher peak. The propulsive peak timing occurred as in walking, but later than running (Figure 5). This could be advantageous to avoid rapid changes in force production but required a constant average activation.

The medio-lateral forces (expressed as delta peaks) in race walking showed the lowest values as in the other gaits, but they were higher than walking and running (Figure 4). This is probably due to the kinematics of the pelvis, which shows a medio-lateral excursion in order to accept the straight knee from heel strike to midstance [22,25,26]. The first lateral peak was simultaneous with the vertical peak, as occurred in walking and running, and the medial peak, which was the greatest, was in line with braking–propulsive inversion.

The BCoM trajectory shows some interesting features of race walking compared with walking and running. At first, the volume was smaller than the other two gaits with less excursion in the medio-lateral direction, which is not expected when considering the great excursion of the pelvis, typical of this gait [22]. However, the BCoM is the weighted mean of all the segments, and a single segment could bring a remarkable bias in the estimated BCoM trajectory [11]. The vertical displacement showed a higher minimum than both walking and running. This can be explained by the race walking rule that requires the knee to be straight during the stance phase. In this phase, BCoM lowers its trajectory as in running, without the knee flexion, race walking BCoM is mechanically forced to stay on a higher trajectory. In walking, when increasing speed, the minimum reaches smaller values. Walking and race walking, at slow speed (2.78 m s$^{-1}$), show the same upper limit in BCoM trajectory. When speed is increased, flight time occurs also in race walking and the maximum is slightly higher, without approaching running values; whereas, the minimum is almost unchanged. The crossing point between the right and left part of the contour in race walking occurs in the upper part as in running, whereas in walking, it is located in the lower part. Also, the potential and kinetic mechanical energies are in phase as in running, whereas in walking, they are out of phase. The symmetry values on the three axes showed a behaviour similar to walking and running. However, a greater number of strides is necessary to give an appropriate description of the symmetry behaviour at increasing speed also in race walking. In conclusion, the BCoM contour of race walking resembles the running pattern, even when no flight time is present, with a smaller excursion.

### 4.3. Different Vertical GRF Groups

Besides the vertical peaks' difference, the clustering of M and N shape athletes showed other small differences in anterior-posterior and medio-lateral forces in the first part of the stance phase, before the braking–propulsion inversion. The braking peak was higher in the N than in the M shape group and the lateral force pattern was less homogeneous with a delayed peak in the N shape group.

Fenton [14] suggested that the magnitude and timing of the vertical peak was an index of smoothness and "fluidity" of the stride. Moreover, the absence of a second vertical peak would direct the force more in the progression direction, on the contrary, a vertical force would cause a vertical displacement that could end in flight phase [14]. In our data, both groups (M and N shape) showed the same first vertical peak, therefore, they should have the same stride "fluidity". As for the vertical displacement, the double integration of acceleration in both groups did not show any appreciable difference in the BCoM trajectory during stance and no difference in flight time.

The different vertical GRF patterns among athletes was already pointed out by Fenton [14], and it was also evident when comparing Fenton's with Cairns and colleagues' [13] data [13,16]. Fenton explained this difference with athletes' performance level: M shape athletes were the least, whereas N shape were the most trained. Despite our relatively small sample size (larger than in the Fenton paper [14]), eight athletes showed an N shape pattern and seven showed an M shape pattern: within groups, the performance level was very different; however, between groups, the level was the same (PB 10,000 m min:s M shape: 44:39; N shape 44:18). This allowed us to conclude that performance level should not be the trigger for different patterns. As shown in Figure 1, the pattern was well characterised and different across the whole range of speeds and no athlete changed it by increasing speed. A further explanation could be related to the different athletes' techniques learnt from different coaches. When analysing this aspect, the sample size decreases even more, with just a couple of athletes for each coach; however, we found that coach technique was not the determinant of the difference either.

## 5. Conclusions

The present comprehensive analysis of ground reaction forces shows that race walking is a gait that shares features with walking and running. Similarly, the increase in speed is achieved by increasing force peaks, which occur at the same relative instant of the stance phase, thus, a comparison among subjects or studies should be done only at the same speed. The peculiarity of race walking kinematics and dynamics features is remarkable, also in the ground reaction forces analysis, since, differently from walking and running, athletes showed two different vertical force patterns within the same gait. These different patterns do not change the 3D trajectory of the body centre of mass and related spatiotemporal parameters, and do not seem to be related to the athletes' performance level. Further investigations are needed to understand which biomechanical factors cause these patterns. The BCoM trajectory obtained by ground reaction forces showed that race walking has the same pattern of running, even at slow speed where no flight time is present, within a smaller volume.

**Author Contributions:** Initial project discussion was conducted by all authors. Specific personal contributions were as follows: conceptualization, G.P.; methodology, G.P., D.C. and A.E.M.; software, G.P., D.C. and A.E.M.; formal analysis, G.P.; investigation, G.P., D.C. and A.L.T.; data curation, G.P.; writing—original draft preparation, G.P.; writing—review and editing, G.P., D.C., A.L.T. and A.E.M.

**Funding:** This research received no external funding.

**Conflicts of Interest:** The authors declare no conflict of interest.

## References


1. Elftman, H. The force exerted by the ground in walking. *Arbeitphysiologie* **1939**, *10*, 485–491. [CrossRef]
2. Fenn, W.O. Work against gravity and work due to velocity changes in running. *Am. J. Physiol.* **1930**, *93*, 433–462. [CrossRef]
3. Muniz, A.M.S.; Nadal, J. Application of principal component analysis in vertical ground reaction force to descriminate normal and abnormal gait. *Gait Posture* **2009**, *29*, 31–35. [CrossRef] [PubMed]
4. White, R.; Agouris, I.; Selbie, R.D.; Kirkpatrick, M. The variability of force platform data in normal and cerebral palsy gait. *Clin. Biomech.* **1999**, *14*, 185–192. [CrossRef]
5. Giakas, G.; Baltzopoulos, V. Time and frequency domain analysis of ground reaction forces during walking: An investigation of variability and symmetry. *Gait Posture* **1997**, *5*, 189–197. [CrossRef]
6. McCrory, L.J.; White, S.C.; Lifeso, R.M. Vertical ground reaction forces: Objetive measures of gait following hip arthroplasty. *Gait Posture* **2001**, *14*, 104–109. [CrossRef]
7. Gerlach, K.E.; White, S.C.; Burton, H.W.; Dorn, J.M.; Leddy, J.J.; Horvath, P.J. Kinetic changes with fatigue and relationship to injury in female runners. *Med. Sci. Sports Exerc.* **2005**, *37*, 657–663. [CrossRef]
8. Anderson, F.C.; Pandy, M.G. Individual muscle contribution to support in normal walking. *Gait Posture* **2003**, *17*, 159–169. [CrossRef]
9. Morin, J.B.; Edouard, P.; Samonzino, P. Technical Ability of Force Application as a Determinant Factor of Sprint Performance. *Med. Sci. Sports Exerc.* **2011**, *43*, 1680–1688. [CrossRef]
10. Cavagna, G.A. Force platforms as ergometers. *J. Appl. Physiol.* **1975**, *39*, 174–179. [CrossRef]
11. Pavei, G.; Seminati, E.; Cazzola, D.; Minetti, A.E. On the Estimation Accuracy of the 3D Body Center of Mass Trajectory during Human Locomotion: Inverse vs. Forward Dynamics. *Front. Physiol.* **2017**, *8*, 129. [CrossRef] [PubMed]
12. Minetti, A.E.; Cisotti, C.; Mian, O.S. The mathematical description of the body centre of mass 3D path in human and animal locomotion. *J. Biomech.* **2011**, *44*, 1471–1477. [CrossRef] [PubMed]
13. Cairns, M.A.; Burdett, R.G.; Pisciotta, J.C.; Simon, S.R. A biomechanical analysis of racewalking gait. *Med. Sci. Sports Exerc.* **1986**, *18*, 446–453. [CrossRef] [PubMed]
14. Fenton, R.M. Race walking ground reaction forces. In *Proceedings of the II International Symposium on Biomechanics in Sports*; Terauds, J., Barthels, K., Krieghbaum, E., Mann, R., Crakes, J., Eds.; Academic: Del Mar, CA, USA, 1984; pp. 61–70.
15. Hanley, B.; Bissas, A. Ground reaction forces of Olympic and World Championship race walkers. *Eur. J. Sport Sci.* **2016**, *16*, 50–56. [CrossRef] [PubMed]
16. Pavei, G.; Cazzola, D.; La Torre, A.; Minetti, A.E. The biomechanics of race walking: Literature overview and new insights. *Eur. J. Sport Sci.* **2014**, *14*, 661–670. [CrossRef] [PubMed]
17. Nilsson, J.; Thorstensson, A. Ground reaction forces at different speeds of human walking and running. *Acta Physiol. Scand.* **1989**, *136*, 217–227. [CrossRef] [PubMed]
18. Saibene, F.; Minetti, A.E. Biomechanical and physiological aspects of legged locomotion in humans. *Eur. J. Appl. Physiol.* **2003**, *88*, 297–316. [CrossRef]
19. Preatoni, E.; La Torre, A.; Rodano, R. A biomechanical comparison between racewalking and normal walking stance phase. In *Proceedings of the XXIV International Symposium on Biomechanics in Sports*; Schwameder, H., Strutzenberger, G., Fastenbauer, V., Lindinger, S., Müller, E., Eds.; Unversitary Press: Salzburg, Austria, 2006.
20. Griffin, T.M.; Kram, R. Penguin waddling is not wasteful. *Nature* **2000**, *408*, 929. [CrossRef]
21. Majed, L.; Heugas, A.-M.; Chamon, M.; Siegler, I.A. Learning an energy-demanding and biomechanically constrained motor skill, racewalking: Movement reorganization and contribution of metabolic efficiency and sensory information. *Hum. Mov. Sci.* **2012**, *31*, 1598–1614. [CrossRef]

22. Pavei, G.; La Torre, A. The effects of speed and performance level on race walking kinematics. *Sport Sci. Health* **2016**, *12*, 35–47. [CrossRef]

23. Alexander, R.M. Mechanics of Bipedal Locomotion. In *Perspective in Experimental Biology 1*; Spencer-Devies, P., Ed.; Pergamon Press: Oxford, UK, 1976; pp. 493–504.

24. Usherwood, J.R.; Channon, A.J.; Myatt, J.P.; Rankin, J.W.; Hubel, T.Y. The human foot and heel-sole-toe walking strategy: A mechanism enabling an inverted pendular gait with low isometric force? *J. R. Soc. Interface* **2012**, *9*, 2396–2402. [CrossRef] [PubMed]

25. Cazzola, D.; Pavei, G.; Preatoni, E. Can coordination variability identify performance factors and skill level in competitive sport? The case of race walking. *J. Sport Health Sci.* **2016**, *5*, 35–43. [CrossRef] [PubMed]

26. Murray, M.P.; Guten, G.; Mollinger, L.; Gardner, G. Kinematic and electromyographic patterns of Olympic race walkers. *Am. J. Sports Med.* **1983**, *11*, 68–74. [CrossRef] [PubMed]

Article

# Core Stability and Symmetry of Youth Female Volleyball Players: A Pilot Study on Anthropometric and Physiological Correlates

Sophia D. Papadopoulou [1], Amalia Zorzou [2], Sotirios Drikos [3], Nikolaos Stavropoulos [1], Beat Knechtle [4] and Pantelis T. Nikolaidis [2,*]

[1] Department of Physical Education & Sport Science, Laboratory of Evaluation of Human Biological Performance, Aristotle University of Thessaloniki, 57001 Thessaloniki, Greece; sophpapa@phed.auth.gr (S.D.P.); nstavrop@phed.auth.gr (N.S.)

[2] Exercise Physiology Laboratory, 18450 Nikaia, Greece; a.zorzou@hotmail.com

[3] School of Physical Education and Sport Science, National and Kapodistrian University of Athens, 17237 Athens, Greece; sodrikos@phed.uoa.gr

[4] Institute of Primary Care, University of Zurich, 8091 Zurich, Switzerland; beat.knechtle@hispeed.ch

* Correspondence: pademil@hotmail.com; Tel.: +30-697-782-0298

Received: 28 January 2020; Accepted: 4 February 2020; Published: 6 February 2020

**Abstract:** The aim of the present study was to examine the variation in core stability and symmetry of youth female volleyball players by age, and its relationship with anthropometric characteristics, the 30 s Wingate anaerobic test (WAnT), and the 30 s Bosco test. Female volleyball players ($n = 24$, age $13.9 \pm 1.9$ years, mean $\pm$ standard deviation) performed a series of anthropometric, core stability tests (isometric muscle endurance of torso flexors, extensors, and right and left lateral bridge), WAnT (peak power, mean power, Pmean, and fatigue index, FI) and Bosco test (Pmean). Flexors-to-extensors ratio and right-to-left lateral bridge ratio were also calculated. Participants were grouped into younger ($n = 12$, $12.3 \pm 1.2$ years) or older than 14 years ($n = 12$, $15.4 \pm 1.0$ years), and into normal (flexors-to-extensors ratio $< 1$; $n = 17$) or abnormal flexors-to-extensors ratio ($\geq 1$; $n = 7$). The older age group was heavier (+11.3 kg, mean difference; 95% CI, 2.0, 20.6) and with higher body mass index (+2.8 kg m$^{-2}$; 95% CI, 0.4, 5.1) than the younger age group. The group with abnormal flexors/extensors had larger flexors muscle endurance (+77.4 s; 95% CI, 41.8, 113.0) and higher flexors/extensors ratio (+0.85; 95% CI, 0.61, 1.10) than the normal group. Body fat percentage (BF) correlated moderately-to-largely with flexors ($r = -0.44$, $p = 0.033$), extensors ($r = -0.51$, $p = 0.011$), and left lateral bridge ($r = -0.45$, $p = 0.027$); WAnT Pmean moderately-to-largely with right ($r = 0.46$, $p = 0.027$) and left lateral bridge ($r = 0.55$, $p = 0.006$); FI moderately-to-largely with right ($r = -0.45$, $p = 0.031$) and left lateral bridge ($r = -0.67$, $p < 0.001$), and right/left ratio ($r = 0.42$, $p = 0.046$); Bosco Pmean correlated moderately-to-largely with right ($r = 0.48$, $p = 0.020$) and left lateral bridge ($r = 0.67$, $p = 0.001$). A stepwise regression analysis indicated FI and BF as the most frequent predictors of core stability. The findings of the present study suggested that increased core stability was related to decreased BF and increased anaerobic capacity. A potential misbalance between torso flexors and extensors might be attributed to bidirectional variations (either high or low scores) of flexors muscle endurance rather than decreased extensors muscle endurance.

**Keywords:** human performance; muscle endurance; team sport; torso extensors; torso flexors

## 1. Introduction

Female volleyball has been one of the most popular team sports worldwide [1]. Performance in this sport has been associated with a series of physical, physiological, psychological and technique and tactical characteristics [2,3]. With regards to physiological characteristics, most studies have focused on jumping ability and anaerobic power so far showing that female volleyball players jumped high and were characterized by high levels of anaerobic power [4,5]. Both jumping ability and anaerobic power might vary by age with adults scoring higher than adolescents [6]. Moreover, they might vary by in-game role of the players, e.g., higher jump height in hitters than libero players [6]. On the other hand, muscle endurance, i.e., the ability maintain muscle power output over time, in female volleyball—despite being a major component of health-related physical fitness—has received less scientific attention [7]. Muscle endurance has been considered not only in terms of absolute values, but also with regards to symmetry between different muscle groups (e.g., agonists versus antagonists) [7].

Core stability, i.e., the ability to optimize the placement and movement of the torso over the pelvis, has been recognized as a major component of muscle endurance. It was observed that core stability was beneficial for human performance, e.g., being stable reference would allow upper and lower limbs developing force [8], and health, e.g., maintenance of low back and knee health [9]. A low level of core stability increased the risk of low back and knee injuries [10]. In volleyball, those with core instability had high scapular malposition, inferior medial border prominence, coracoid pain, and dyskinesis of scapular movement [11]. Furthermore, the inclusion of core stability exercises was considered in preventive training programs [12,13]. With regards to the symmetry of the muscle endurance of torso muscles, e.g., flexors-to-extensors or right-to-left lateral flexors, few studies were conducted in sports [14,15] including volleyball [7]. It has been proposed that a ratio of torso flexors-to-extensors muscle endurance larger than one might indicate misbalance in the torso muscle groups [16], and consequently, this ratio could be used in volleyball to monitor muscle imbalances and identify potential injury risk. Volleyball included overhead tasks relied on shoulder movements, which in turn needed core stability to be efficient [7], and it has been shown that core stability might influence muscle strength of shoulders [17].

Although the abovementioned studies improved our understanding about the role of core stability and symmetry on health, little information existed about its role on performance in volleyball. The knowledge of the relationship of core stability and symmetry with anthropometric and physiological characteristics would be of practical value for professionals working with female volleyball players. In addition to the symmetry of core muscle endurance, it would be also interesting to examine the metabolic aspect, where it might be assumed that it would rely on the anaerobic energy transfer system considering its duration (several seconds) and exercise intensity (increased muscle activity) [18]. In exercise testing, the Wingate anaerobic test (WAnT) has been considered as a "golden" standard of anaerobic power and capacity despite its specific mode of exercise (cycling) [19]. A continuous 30 s Bosco jumping test has been developed as more sport-specific than WAnT to monitor performance, especially in sports involving many jumps [20,21]. Therefore, information on the relationship of core stability and symmetry with WAnT and Bosco test would provide insight into the metabolic demands of exercise testing of the former variables. In turn, anaerobic capacity has been shown to be inversely related with body fat percentage (BF) [22], i.e., the higher the BF, the lower the anaerobic capacity, and, consequently, it might be expected that BF would be related with core stability indices too.

With regards to correlates of core stability with physiological measures, research on female soccer players reported no correlation of core stability with sprint and muscle strength; however, this finding might be due to the sample size of this study [23]. In addition, information about the variation of core stability and symmetry by age in female volleyball would also be interesting in terms of training and testing. It has been shown that the prevalence of back pain was higher in 14–17 than 11–13 year-old athletes [24]. Moreover, anaerobic capacity assessed by the WAnT and Bosco test was larger in 14–18 than in under 14 year-old female volleyball players [25], whereas no difference was observed in sit-ups test between under and over 14 years female volleyball players [26]. However, no information on age related differences in volleyball has been examined previously with regard to core stability and symmetry. Therefore, the aim of the present study was to examine the variation in core stability and symmetry of female volleyball players by age and its relationship with anthropometric characteristics, WAnT, and the Bosco test. A secondary aim was to compare examine differences between groups varying for torso flexors-to-extensors ratio as it was suggested that a ratio $\geq 1$ would indicate misbalance [16]. The research hypothesis was that increased core stability indices would be associated with high scores of WAnT and Bosco test indices, and low BF. Since anaerobic capacity and body composition were related to performance [4,5], a potential association of core stability indices with these variables would highlight the relevance of core stability with performance.

## 2. Materials and Methods

### 2.1. Study Design and Participants

Female volleyball players ($n = 24$, age $13.9 \pm 1.9$ years) performed a series of anthropometric, core stability (isometric muscle endurance of torso flexors, extensors, and right and left lateral bridge) and WAnT (peak power, mean power, Pmean, and fatigue index, FI, were estimated). Since there was no information about minimal level of effect size in the differences between groups that would be of scientific interest, the sample size was selected considering previous studies [7,23]. Flexors-to-extensors ratio and right-to-left lateral bridge ratio were also calculated to evaluate the symmetry of the core stability variables. Participants were volleyball players of a sport club in Athens and volunteered for this study. They had sport experience $2.9 \pm 1.9$ years, practiced volleyball $3.7 \pm 1.0$ days per week with each training session lasting $93 \pm 10$ min, a total weekly training volume $348 \pm 124$ min and participated in one official game per week. After being informed with details about all procedures, participants and their guardians provided their consent to participate. The exercise testing was performed in a single session. The study was approved by the local Committee of Ethics (EPL 2019/12). Participants were grouped into younger ($n = 12$, age $12.3 \pm 1.2$ years, sport experience $2.5 \pm 1.6$ years, $3.6 \pm 0.6$ weekly training units, and volume $321 \pm 54$ min) or older than 14 years ($n = 12$, $15.4 \pm 1.0$ years, $3.4 \pm 2.1$ years, $3.8 \pm 1.2$ and $374 \pm 166$ min, respectively), and into normal (flexors-to-extensors ratio $< 1$; $n = 17$) or abnormal flexors-to-extensors ratio ($\geq 1$; $n = 7$) according to the classification of McGill [16]. An age of 14 years has been suggested to categorize pubertal status in girls [27] and classified adolescent female volleyball players into age groups [25,26]. Considering the sample size and their small sport experience, the participants were not grouped by playing position.

### 2.2. Equipment and Procedures

Participants were evaluated for stature (SECA, Leicester, UK) and body mass (HD-351 Tanita, City, IL, USA) to the nearest 0.1 cm and 0.1 kg, respectively. The thickness of ten skinfolds (cheek, chin, pectoral, triceps, subscapular, abdomen, chest II, iliac crest, patella and proximal calf) was measured on the right side of the body to the nearest 0.1 mm (Harpenden, West Sussex, UK) and was used to estimate BF according to a Parizkova equation described by Eston and Reilly [28]. After a standardized warm-up including 9 min submaximal cycling and 6 min stretching exercises, participants performed the 30 s Wingate anaerobic test (WAnT) on a cycle ergometer (874 Ergomedic, Monark, City, Sweden) against braking force $0.075 \times$ body mass providing peak power (Ppeak, W kg$^{-1}$), mean power (Pmean, W kg$^{-1}$), and fatigue index (FI, %). Participants were informed that WAnT was an all-out test not allowing the adoption of a pacing strategy, and were encouraged continuously during the test to exert maximal effort. In addition, a continuous 30 s jumping Bosco test was performed, where the participants were instructed to jump continuously throughout this period aiming to achieve maximal jump height in each jump, minimal time spent at the ground between consecutive jumps and maintaining their hands on the hips [20]. The mean power (Ppeak, W kg$^{-1}$) was the outcome measure of the Bosco test.

To assess core stability, four primary (torso flexors, extensors, right and left lateral bridge test) and two secondary measures (flexors to extensors ratio and right to left lateral ratio) following the recommendations of Hoogenboom and Bennett [29] were performed. Participants were familiarized with these measures, since they were included in their training routine. In the torso flexors test, the participant adopted a sit-up position at angle 60° from the floor, whereas, in the torso extensors test, the participant was with her upper body unsupported out of a table and an ankle 180° at hips. In the lateral bridge test, the participant was lying using a side-bridge position. A few seconds practice was provided prior to testing to explain the correct position. A single trial was performed for each test and a 5-min break was provided between tests to allow sufficient recovery [18]. In each test of core stability, participants were asked to maintain the correct position as much as possible. Each primary measure was evaluated in the nearest 0.1 s; thereafter, the secondary measures were calculated to the nearest 0.1. The timing of each test started when participants adopted the instructed position and stopped when a deviation from the position was observed. This protocol evaluated core stability and symmetry previously in female and male soccer players [14,15]. Reliability coefficients ranged from 0.93 (flexors), and 0.96 (right later bridge) to 0.99 (extensors and left lateral bridge) [18].

*2.3. Statistical and Data Analyses*

IBM SPSS v.23.0 (SPSS, Chicago, USA) and Graphpad v.7.0 (GraphPad Prism, San Francisco, CA, USA) were used for statistical analyses. Although the data did not present normal distribution according to visual inspection of Q–Q plots and Shapiro–Wilk test (since $n$ was lower than 50), parametric statistics were used to provide comparable methods and analysis with previous studies on core stability [16,18,23,29]. A non-parametric statistics (median, inter-quartile range, Mann–Whitney U test for differences between groups and Spearman rho for correlations among variables) were also presented in Tables 1–3 to maintain the statistical integrity of this paper. Data were expressed as mean and standard deviation. A preliminary examination of potential relationship of training characteristics with the variables of interest did not reveal any significant correlation; thus, training characteristics were not considered as covariate. An independent student *t*-test examined differences between age groups (under 14 years versus over 14 years) and torso flexors-to-extensors ratio groups (normal versus abnormal). The magnitude of these differences was evaluated by Cohen's d, classified as trivial ($d \leq 0.2$), small ($0.2 < d \leq 0.6$), moderate ($0.6 < d \leq 1.2$), large ($1.2 < d \leq 2.0$), or very large ($d > 2.0$) [30]. The relationship of core stability and symmetry (torso flexors, extensors, right and left lateral bridge test, flexors to extensors ratio, and right to left lateral ratio) with anthropometric characteristics (age, height, weight, body mass index and BF), WAnT (Ppeak, Pmean and FI), and Bosco test (Pmean) was examined using Pearson correlation r. A step-wise regression analysis examined predictors of core stability and symmetry. Statistical significance was set at alpha 0.05.

**Table 1.** Descriptive statistics by age group.

| Variable | Under 14 Years ($n = 12$) | | Over 14 Years ($n = 12$) | |
|---|---|---|---|---|
| | Mean ± SD | Median (IQR) | Mean ± SD | Median (IQR) |
| Anthropometry | | | | |
| Age (years) | 12.3 ± 1.2 | 12.5 (11.1–13.4) | 15.4 ± 1.0 ** | 15.5 (14.6–15.9) ** |
| Body mass (kg) | 53.3 ± 8.5 | 54.7 (44.7–60.8) | 64.6 ± 12.9 * | 62.4 (53.6–75.4) |
| Height (m) | 1.60 ± 0.08 | 1.60 (1.53–1.66) | 1.66 ± 0.07 | 1.68 (1.60–1.72) |
| BMI (kg.m$^{-2}$) | 20.6 ± 2.3 | 20.8 (18.1–22.5) | 23.4 ± 3.1 * | 22.8 (20.4–25.8) * |
| BF (%) | 23.8 ± 5.4 | 23.8 (20.1–28.7) | 25.0 ± 3.5 | 25.3 (23.2–27.7) |
| Core stability | | | | |
| Flexors (s) | 86.8 ± 69.6 | 55.5 (33.8–138.9) | 66.8 ± 23.8 | 64.9 (46.5–80.0) |
| Extensors (s) | 107.6 ± 50.0 | 96.2 (71.7–138.9) | 109.6 ± 25.1 | 107.4 (92.3–122.6) |
| Right lateral (s) | 31.3 ± 17.5 | 34.6 (11.5–48.9) | 35.0 ± 13.3 | 35.6 (20.9–45.4) |
| Left lateral (s) | 34.8 ± 16.3 | 35.1 (22.4–51.0) | 42.7 ± 16.8 | 44.4 (33.6–48.0) |
| Flexors/extensors | 0.84 ± 0.60 | 0.65 (0.31–1.22) | 0.64 ± 0.28 | 0.52 (0.43–0.79) |
| Right/left lateral | 0.90 ± 0.28 | 0.89 (0.77–1.08) | 0.92 ± 0.41 | 0.85 (0.61–1.16) |

**Table 1.** *Cont.*

| Variable | Under 14 Years (*n* = 12) | | Over 14 Years (*n* = 12) | |
|---|---|---|---|---|
| | Mean ± SD | Median (IQR) | Mean ± SD | Median (IQR) |
| **WAnT** | | | | |
| Ppeak (W kg$^{-1}$) | 8.28 ± 0.81 | 8.35 (7.69–8.84) | 8.85 ± 0.47 | 8.86 (8.72–9.13) * |
| Pmean (W kg$^{-1}$) | 5.94 ± 0.78 | 5.99 (5.13–6.59) | 6.57 ± 0.84 | 6.74 (6.00–6.86) |
| FI (%) | 49.9 ± 8.0 | 50.7 (42.6–55.0) | 45.9 ± 7.1 | 45.5 (41.3–46.6) |
| **Bosco test** | | | | |
| Pmean (W kg$^{-1}$) | 24.1 ± 4.1 | 23.7 (20.8–26.4) | 25.4 ± 4.1 | 24.8 (23.1–26.5) |

SD = standard deviation, IQR = inter-quartile range, BMI = body mass index, BF = body fat percentage, flexors-to-extensors ratio, right-to-left lateral bridge ratio, WAnT = Wingate anaerobic test, Ppeak = peak power, Pmean = mean power, FI = fatigue index; * $p < 0.05$, ** $p < 0.001$.

**Table 2.** Descriptive statistics by flexors-to-extensors ratio group.

| Variable | Normal Flexors-to-Extensors (*n* = 17) | | Abnormal Flexors-to-Extensors (*n* = 7) | |
|---|---|---|---|---|
| | Mean ± SD | Median (IQR) | Mean ± SD | Median (IQR) |
| **Anthropometry** | | | | |
| Age (years) | 13.9 ± 2.1 | 14.1 (11.9–15.6) | 13.7 ± 1.3 | 13.7 (13.0–14.6) |
| Body mass (kg) | 60.1 ± 13.3 | 56.1 (53.3–71.4) | 56.2 ± 9.0 | 60.1 (44.8–62.2) |
| Height (m) | 1.63 ± 0.08 | 1.64 (1.57–1.70) | 1.63 ± 0.09 | 1.65 (1.53–1.67) |
| BMI (kg.m$^{-2}$) | 22.4 ± 3.3 | 22.4 (19.8–24.9) | 21.0 ± 2.1 | 22.1 (19.0–22.5) |
| BF (%) | 25.2 ± 4.7 | 25.7 (23.4–28.9) | 22.3 ± 3.2 | 22.6 (20.0–24.5) |
| **Core stability** | | | | |
| Flexors (s) | 54.2 ± 20.6 | 53.8 (41.0–72.9) | 131.6 ± 65.0 * | 123.9 (68.2–193.2) ** |
| Extensors (s) | 114.1 ± 39.8 | 107.9 (84.4–133.3) | 95.2 ± 35.0 | 102.1 (62.2–107.7) |
| Right lateral (s) | 32.9 ± 14.3 | 30.4 (18.2–44.9) | 33.8 ± 18.8 | 39.9 (9.0–50.6) |
| Left lateral (s) | 36.8 ± 13.8 | 38.8 (29.4–48.8) | 43.5 ± 23.0 | 39.5 (32.5–56.6) |
| Flexors/extensors | 0.49 ± 0.18 | 0.46 (0.36–0.65) | 1.35 ± 0.40 * | 1.17 (1.10–1.64) * |
| Right/left bridge | 0.95 ± 0.36 | 0.89 (0.73–1.20) | 0.82 ± 0.30 | 0.89 (0.65–1.10) |
| **WAnT** | | | | |
| Ppeak (W kg$^{-1}$) | 8.55 ± 0.69 | 8.78 (8.31–8.94) | 8.59 ± 0.82 | 8.81 (7.56–9.17) |
| Pmean (W kg$^{-1}$) | 6.13 ± 0.89 | 6.46 (5.22–6.80) | 6.61 ± 0.72 | 6.55 (5.99–6.99) |
| FI (%) | 48.7 ± 7.9 | 46.4 (44.3–53.4) | 45.8 ± 7.1 | 44.9 (40.4–55.0) |
| **Bosco test** | | | | |
| Pmean (W kg$^{-1}$) | 24.1 ± 4.2 | 24.3 (20.2–26.4) | 26.1 ± 3.7 | 25.4 (23.6–26.5) |

IQR = inter-quartile range, BMI = body mass index, BF = body fat percentage, flexors-to-extensors ratio, right-to-left lateral bridge ratio, WAnT = Wingate anaerobic test, Ppeak = peak power, Pmean = mean power, FI = fatigue index; * $p < 0.001$, ** $p < 0.01$.

**Table 3.** Correlations r (Spearman rho in brackets) of core stability and symmetry indices with anthropometric characteristics and Wingate anaerobic test.

| Variable | Core Stability and Symmetry | | | | | |
|---|---|---|---|---|---|---|
| | Flexors (s) | Extensors (s) | Right Lateral (s) | Left Lateral (s) | Flexors/Extensors | Right/Left |
| Age (years) | −0.06 (0.06) | −0.21 (−0.05) | −0.05 (−0.06) | 0.12 (0.12) | 0.06 (0.10) | −0.11 (−0.13) |
| Body mass (kg) | −0.23 (−0.13) | −0.32 (−0.29) | −0.27 (−0.31) | −0.21 (−0.17) | −0.11 (0.03) | 0.18 (−0.10) |
| Height (m) | −0.06 (0.04) | −0.26 (−0.12) | −0.19 (−0.20) | −0.04 (−0.05) | <0.01 (0.14) | −0.02 (−0.10) |
| BMI (kg m$^{-2}$) | −0.31 (−0.18) | −0.35 (−0.34) | −0.30 (−0.27) | −0.30 (−0.23) | −0.16 (−0.02) | 0.24 (0.04) |
| BF (%) | −0.44 * (−0.46 *) | −0.51 * (−0.26) | −0.36 (−0.34) | −0.45 * (−0.53 **) | −0.27 (−0.33) | 0.18 (0.16) |
| Ppeak (W kg$^{-1}$) | −0.20 (0.03) | −0.10 (−0.14) | −0.01 (−0.11) | 0.04 (−0.06) | −0.09 (0.10) | −0.01 (−0.18) |

**Table 3.** *Cont.*

| | Core Stability and Symmetry | | | | | |
|---|---|---|---|---|---|---|
| Variable | Flexors (s) | Extensors (s) | Right Lateral (s) | Left Lateral (s) | Flexors/Extensors | Right/Left |
| WAnT Pmean (W kg$^{-1}$) | 0.29 (0.49 *) | 0.19 (0.12) | 0.46 * (0.39) | 0.55 ** (0.53 *) | 0.30 (0.39) | −0.13 (−0.18) |
| FI (%) | −0.39 (−0.54 **) | −0.14 (−0.07) | −0.45 * (−0.41) | −0.67 *** (−0.66 **) | −0.39 (−0.41) | 0.42 * (0.33) |
| Bosco Pmean (W kg$^{-1}$) | 0.25 (0.45 *) | 0.24 (0.16) | 0.48 * (0.50 *) | 0.67 ** (0.63 **) | 0.19 (0.32) | −0.20 (−0.07) |

BMI = body mass index, BF = body fat percentage, flexors-to-extensors ratio, right-to-left lateral bridge ratio, Ppeak = peak power, WAnT = Wingate anaerobic test, Pmean = mean power, FI = fatigue index; * $p < 0.05$, ** $p < 0.01$, *** $p < 0.001$.

## 3. Results

The older age group differed in age from the younger one by 3.1 years (95% confidence intervals, CI, 2.2, 4.0; Cohen's $d = 2.8$), was heavier (+11.3 kg, mean difference; 95% CI, 2.0, 20.6; $d = 1.0$) and had a higher body mass index (+2.8 kg m$^{-2}$; 95% CI, 0.4, 5.1; $d = 1.0$) (Table 1). No other difference was observed between age groups ($p > 0.05$). The group with abnormal flexors/extensors had larger flexors muscle endurance (+77.4 s; 95% CI, 41.8, 113.0; $d = 1.6$) and lower flexors/extensors (+0.85; 95% CI, 0.61, 1.10; $d = 2.8$) than the normal group (Table 2). No other difference was shown between flexors/extensors groups ($p > 0.05$).

The correlations of core stability and symmetry indices with anthropometric characteristics, WAnT and Bosco test were presented in Table 3. BF correlated moderately-to-largely with flexors, extensors and left lateral bridge, WAnT and Bosco Pmean moderately-to-largely with right and left lateral bridge, and FI moderately-to-largely with right and left lateral bridge, and right/left ratio. Representative correlations were depicted in Figure 1. The findings of the stepwise regression analysis were presented in Table 4. FI and BF were the most frequent predictors of core stability and symmetry.

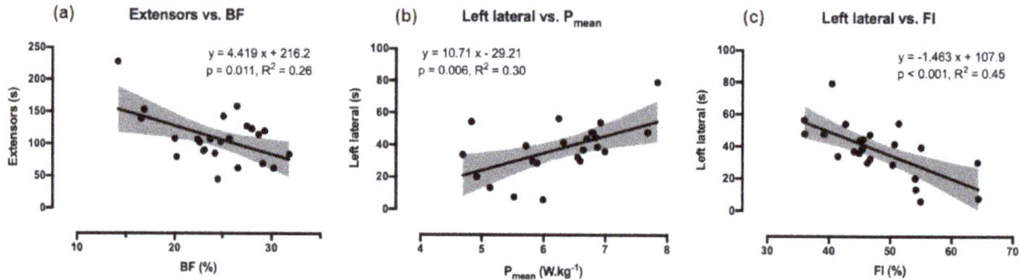


**Figure 1.** Relationship of core stability with body fat percentage (**BF**; **a**), mean power (**Pmean**; **b**), and fatigue index (**FI**; **c**) of the Wingate anaerobic test.

**Table 4.** Stepwise regression analysis.

| | Core Stability and Symmetry | | | | | |
|---|---|---|---|---|---|---|
| Variable | Flexors (s) | Extensors (s) | Right Lateral (s) | Left Lateral (s) | Flexors/Extensors | Right/Left |
| Predictors | BF, FI | BF | WAnT Pmean, Ppeak | FI, Bosco Pmean | BF, FI | FI |
| R | 0.75 | 0.51 | 0.73 | 0.85 | 0.71 | 0.42 |
| R2 | 0.56 | 0.26 | 0.54 | 0.72 | 0.50 | 0.18 |
| SEE | 37.9 | 34.0 | 11.1 | 9.3 | 0.35 | 0.33 |

BF = body fat percentage, flexors-to-extensors ratio, right-to-left lateral bridge ratio, Ppeak = peak power, Pmean = mean power, FI = fatigue index, WAnT = Wingate anaerobic test; SEE = standard error of the estimate.

## 4. Discussion

The main findings of the present study were that (a) no difference in core stability and symmetry was observed between age groups; (b) participants with abnormal flexors-to-extensors ratio had more muscle endurance in flexors than those with normal ratio; (c) flexors and extensors muscle endurance correlated with BF, i.e., the larger the muscle endurance, the lower the BF; (d) lateral muscle endurance correlated with indices of WAnT and Bosco test; and (e) FI and BF were the most frequent predictor of core stability and symmetry.

Considering the role of age, the comparison between age groups (~12 versus ~15 years) did not show any difference in core stability and symmetry, which was in agreement with a study on adolescent non-athletes [31]. It might be assumed that the intermittent nature of volleyball did not facilitate the development of muscle endurance. The relationship of core stability with BF might be attributed to the negative role of BF in exercise performance related to muscle endurance and anaerobic capacity [22,32]. Previously, it was observed that BF correlated with WAnT Pmean in both adolescent and adult female volleyball players [22], where high BF was related to low WAnT Pmean. It has been also shown that a higher BF was related to a lower number of sit-ups in 1 min in female police officers [32]. This negative role of BF for core stability and muscle endurance might be that fat was an extra load that should be sustained without contributing to muscle contraction.

Core stability indices (lateral bridge) correlated either with WAnT Pmean, i.e., mean cycling performance over 30s, or FI, i.e., percentage decrease of cycling performance over 30 s. Particularly, a high score of lateral bridge was related to high score of Pmean and low score of FI. It should be highlighted that a low score of FI in WAnT-combined with adequate Pmean-indicated high anaerobic capacity, since a participant was able to maintain performance during prolonged exercise [33]. On the other hand, no correlation was observed between core stability and Ppeak, i.e., performance in the first 5 s of WAnT. This finding was in agreement with a study in female and male soccer players, where core stability did not correlate with isometric muscle strength [15]. From a physiological point of view, core stability tests lasting from 6 s to 230 s had closer affinity with anaerobic capacity (WAnT Pmean and FI) rather than muscle power (Ppeak) and muscle strength. Moreover, it has been observed that exercise duration might partially explain the similar results of two different modes (cycling versus jumping) of exercise tests [21]. With regard to the correlations of the Bosco test, it was observed that the performance on this test correlated moderately-to-largely with torso lateral flexors muscle endurance. This observation was in agreement with research showing that torso lateral flexors had substantial potentials as stabilizers and energy generators during jumps [34] and played an important role in single-leg jumps independently of vertical or horizontal direction [35].

With regard to torso flexors-to-extensors ratio, it was found that an increased ratio in the abnormal group was due to an increased score of flexors. A ratio of 1.15 was observed in workers with a history of back disorders compared to 0.71 in their healthy counterparts [16], where the 1.15 ratio was attributed more to weak extensors rather than to strong flexors. These findings implied that, although the abnormal group of volleyball players had increased flexors-to-extensors ratio–observed also in a group with history of back disorders [16]—a different aetiology might be assumed (increased muscle endurance of flexors in the former group versus decreased muscle endurance of torso extensors in the latter group). The overall torso flexors-to-extensors ratio in the present study (0.74) was similar to that of healthy adults (0.71) [16] and adult female volleyball players (0.73) [7]. It should be highlighted that, although our results about torso flexors-to-extensors ratio were similar to their adult counterparts [7], the absolute scores of torso flexors and extensors muscle endurance were quite lower in our sample. Thus, the higher values of torso flexors and extensors in adult female volleyball players [7] might be attributed to a long-term training effect. Moreover, low back pain was associated with torso extensors and flexors weakness [36]. With regard to the normal group, i.e., the group with torso flexors-to-extensors ratio lower than one, it was observed that this ratio (0.49) was lower than that reported by literature on healthy adults and adult female volleyball players (~0.72) [7,16]. It was also shown that this decreased ratio was attributed more to decreased flexors muscle endurance rather than to the score of extensors. Thus, a training aim should be to prevent torso flexor-to-extensors misbalance in both directions (i.e., low or high scores of flexors muscle endurance.

The mean score of right-to-left lateral bridge (≥0.90) indicated a relative symmetry between right and left side of the torso; however, the large variation of scores of the right-to-left lateral bridge ratio shown by SD 0.28–0.41 suggested a lack of symmetry between the two sides, i.e., there were participants with large differences in muscle endurance between right and left torso flexors. On the other hand, SD as a measure of inter-individual variation should be interpreted with caution considering the lack of normal distribution of the data as indicated by non-parametric statistics (Tables 1 and 2). These findings highlighted the need for a balanced training load between torso flexors and extensors as well as between right and left lateral muscle groups.

A limitation of the present study was that the exercise tests of core stability relied on isometric muscle contraction; thus, caution would be needed to generalize the findings to exercise tests using other modes of muscle contraction (isotonic or isokinetic). Furthermore, special attention would be necessary when performing exercise tests of core stability such as torso flexors, since even a minimal deviation from the correct position would result in altered muscle activation influencing the outcome [37]. It was also acknowledged that other assessment methods of muscle symmetry (e.g., surface electromyography and isokinetic dynamometry) should be selected in future studies to verify our findings by using laboratory methods. On the other hand, the strength of this study was its novelty as it provided evidence about the relationship of core stability indices with BF, WAnT, and the Bosco test. In addition, the findings had practical applications for physicians, exercise physiologists, and fitness trainers to monitor the training of volleyball players. Performance in volleyball relied on the effectiveness of the dynamic movements of the shoulder, which in turn performed movements taking advantage of a stable torso; in this sense, although the core did not participate directly in dynamic movements, its optimal muscle function was necessary to stabilize the shoulder zone and pelvis in order for upper and lower limbs to perform efficiently [7]. Since data on core stability of adolescent female volleyball players were not available in the existed literature, practitioners might use our findings as reference to evaluate core stability and symmetry of their athletes. Although a correlation would not indicate causation, the knowledge of the relationship of core stability with BF, WAnT, and the Bosco test would aid practitioners in the interpretation of core stability measurements, e.g., a low score of core stability of an athlete with high BF might be attributed to an excess BF in addition to a likely muscle weakness.

## 5. Conclusions

The findings of the present study suggested that increased core stability was related to decreased BF and increased anaerobic capacity. A potential misbalance between torso flexors and extensors might be attributed to bidirectional variations (either high or low scores) of flexors rather than decreased extensors muscle endurance. Considering the lack of available data on core stability and symmetry in adolescent female volleyball players, our findings could be used by practitioners in the context of testing and training.

**Author Contributions:** Conceptualization, S.D.P. and P.T.N.; methodology, S.D.P. and P.T.N.; software, P.T.N.; validation, S.D.P. and P.T.N.; formal analysis, S.D.P. and P.T.N.; investigation, S.D.P. and P.T.N.; resources, S.D.P. and P.T.N.; data curation, S.D.P. and P.T.N.; writing—original draft preparation, S.D.P., A.Z., S.D., N.S., B.K., and P.T.N.; writing—review and editing, S.D.P., A.Z., S.D., N.S., B.K., and P.T.N.; visualization, S.D.P. and P.T.N.; supervision, S.D.P., B.K., and P.T.N.; project administration, S.D.P., B.K., and P.T.N. All authors have read and agreed to the published version of the manuscript.

**Funding:** This research received no external funding.

**Acknowledgments:** The voluntary participation of volleyball players in this research and the collaboration with the technical staff were gratefully acknowledged.

**Conflicts of Interest:** The authors declare no conflict of interest.

## References


1. Greco, G.; Messina, G.; Angiulli, A.; Patti, A.; Iovane, A.; Fischetti, F. A preliminary comparative study on the effects of pilates training on physical fitness of young female volleyball players. *Acta Med. Mediterr.* **2019**, *35*, 783–789. [CrossRef]
2. Rikberg, A.; Raudsepp, L. Multidimensional performance characteristics in talented male youth volleyball players. *Pediatr. Exerc. Sci.* **2011**, *23*, 537–548. [CrossRef] [PubMed]
3. Vargas, J.; Loureiro, M.; Nikolaidis, P.T.; Knechtle, B.; Laporta, L.; Marcelino, R.; Afonso, J. Rethinking Monolithic Pathways to Success and Talent Identification: The Case of the Women's Japanese Volleyball Team and Why Height is Not Everything. *J. Hum. Kinet.* **2018**, *64*, 233–245. [CrossRef] [PubMed]
4. Kavanaugh, A.A.; Mizuguchi, S.; Sands, W.A.; Ramsey, M.W.; Stone, M.H. Long-Term changes in jump performance and maximum strength in a cohort of national collegiate athletic association division I women's volleyball athletes. *J. Strength Cond. Res.* **2018**, *32*, 66–75. [CrossRef]
5. Nikolaidis, P.T.; Afonso, J.; Busko, K. Differences in anthropometry, somatotype, body composition and physiological characteristics of female volleyball players by competition level. *Sport Sci. Health* **2015**, *11*, 29–35. [CrossRef]
6. Nikolaidis, P.T.; Afonso, J.; Buśko, K.; Ingebrigtsen, J.; Chtourou, H.; Martin, J.J. Positional differences of physical traits and physiological characteristics in female volleyball players-The role of age. *Kinesiology* **2015**, *47*, 75–81.
7. Zandi, S.; Rajabi, R.; Minoonejad, H.; Mohseni-Bandpei, M. Core muscular endurance in volleyball players with anterior shoulder instability and asymptomatic players. *Med. Dello Sport* **2018**, *71*, 96–106. [CrossRef]
8. Willardson, J.M. Core stability training: Applications to sports conditioning programs. *J. Strength Cond. Res.* **2007**, *21*, 979–985. [CrossRef]
9. Willson, J.D.; Dougherty, C.P.; Ireland, M.L.; Davis, I.M. Core stability and its relationship to lower extremity function and injury. *J. Am. Acad. Orthop. Surg.* **2005**, *13*, 316–325. [CrossRef]
10. Borghuis, J.; Hof, A.L.; Lemmink, K.A. The importance of sensory-Motor control in providing core stability: Implications for measurement and training. *Sports Med. (Auckland N.Z.)* **2008**, *38*, 893–916. [CrossRef]
11. Reeser, J.C.; Joy, E.A.; Porucznik, C.A.; Berg, R.L.; Colliver, E.B.; Willick, S.E. Risk factors for volleyball-Related shoulder pain and dysfunction. *PM&R* **2010**, *2*, 27–36. [CrossRef]
12. Leporace, G.; Praxedes, J.; Pereira, G.R.; Pinto, S.M.; Chagas, D.; Metsavaht, L.; Chame, F.; Batista, L.A. Influence of a preventive training program on lower limb kinematics and vertical jump height of male volleyball athletes. *Phys. Ther. Sport Off. J. Assoc. Chart. Physiother. Sports Med.* **2013**, *14*, 35–43. [CrossRef] [PubMed]
13. Gouttebarge, V.; van Sluis, M.; Verhagen, E.; Zwerver, J. The prevention of musculoskeletal injuries in volleyball: The systematic development of an intervention and its feasibility. *Inj. Epidemiol.* **2017**, *4*, 25. [CrossRef] [PubMed]
14. Nikolaidis, P.T. Core stability of male and female football players. *Biomed. Hum. Kinet.* **2010**, *2*, 30–33. [CrossRef]
15. Roth, R.; Donath, L.; Zahner, L.; Faude, O. Muscle activation and performance during trunk strength testing in high-Level female and male football players. *J. Appl. Biomech.* **2016**, *32*, 241–247. [CrossRef]
16. McGill, S.M. *Ultimate Back Fitness and Performance*; Wabuno Publishers: Waterloo, ON, Canada, 2004.
17. Rosemeyer, J.R.; Hayes, B.T.; Switzler, C.L.; Hicks-Little, C.A. Effects of Core-Musculature fatigue on maximal shoulder strength. *J. Sport Rehabil.* **2015**, *24*, 384–390. [CrossRef]
18. McGill, S.M.; Childs, A.; Liebenson, C. Endurance times for low back stabilization exercises: Clinical targets for testing and training from a normal database. *Arch. Phys. Med. Rehabil.* **1999**, *80*, 941–944. [CrossRef]
19. Yapici, A.; Findikoglu, G.; Dundar, U. Do isokinetic angular velocity and contraction types affect the predictors of different anaerobic power tests? *J. Sports Med. Phys. Fit.* **2016**, *56*, 383–391.
20. Čular, D.; Ivančev, V.; Zagatto, A.M.; Milić, M.; Beslija, T.; Sellami, M.; Padulo, J. Validity and Reliability of the 30-s Continuous Jump for Anaerobic Power and Capacity Assessment in Combat Sport. *Front. Physiol.* **2018**, *9*, 543. [CrossRef]
21. Nikolaidis, P.T.; Afonso, J.; Clemente-Suarez, V.J.; Alvarado, J.R.P.; Driss, T.; Knechtle, B.; Torres-Luque, G. Vertical Jumping Tests versus Wingate Anaerobic Test in Female Volleyball Players: The Role of Age. *Sports* **2016**, *4*, 9. [CrossRef]

22. Nikolaidis, P.T. Body mass index and body fat percentage are associated with decreased physical fitness in adolescent and adult female volleyball players. *J. Res. Med Sci. Off. J. Isfahan Univ. Med Sci.* **2013**, *18*, 22–26.

23. Nesser, T.W.; Lee, W.L. The relationship between core strength and performance in division I female soccer players. *J. Exerc. Physiol. Online* **2009**, *12*, 21–28.

24. Müller, J.; Müller, S.; Stoll, J.; Fröhlich, K.; Otto, C.; Mayer, F. Back pain prevalence in adolescent athletes. *Scand. J. Med. Sci. Sports* **2017**, *27*, 448–454. [CrossRef] [PubMed]

25. Nikolaidis, P.T.; Ziv, G.; Arnon, M.; Lidor, R. Physical characteristics and physiological attributes of female volleyball players-The need for individual data. *J. Strength Cond. Res.* **2012**, *26*, 2547–2557. [CrossRef] [PubMed]

26. Melrose, D.R.; Spaniol, F.J.; Bohling, M.E.; Bonnette, R.A. Physiological and performance characteristics of adolescent club volleyball players. *J. Strength Cond. Res.* **2007**, *21*, 481–486. [CrossRef] [PubMed]

27. Gelbart, M.; Ziv-Baran, T.; Williams, C.A.; Yarom, Y.; Dubnov-Raz, G. Prediction of Maximal Heart Rate in Children and Adolescents. *Clin. J. Sport Med. Off. J. Can. Acad. Sport Med.* **2017**, *27*, 139–144. [CrossRef] [PubMed]

28. Eston, R.; Reilly, T. *Kinanthropometry and Exercise Physiology Laboratory Manual. Tests, Procedures and Data: Volume 1: Anthropometry*, 3rd ed.; Routledge: London, UK, 2009; pp. 32–35.

29. Hoogenboom, B.J.; Bennett, J.L. Core stabilization training in rehabilitation. In *Musculoskeletal Interventions, Techniques for Therapeutic Exercise*; Voight, M.L., Hoogenboom, B.J., Prentice, W.E., Eds.; McGraw-Hill: New York, NY, USA, 2007; pp. 333–358.

30. Cohen, J. *Statistical Power Analysis for the Behavioral Sciences*; Lawrence Erlbaum Associates: Hillsdale, NJ, USA, 1988.

31. Dejanovic, A.; Harvey, E.P.; McGill, S.M. Changes in torso muscle endurance profiles in children aged 7 to 14 years: Reference values. *Arch. Phys. Med. Rehabil.* **2012**, *93*, 2295–2301. [CrossRef]

32. Violanti, J.M.; Ma, C.C.; Fekedulegn, D.; Andrew, M.E.; Gu, J.K.; Hartley, T.A.; Charles, L.E.; Burchfiel, C.M. Associations Between Body Fat Percentage and Fitness among Police Officers: A Statewide Study. *Saf. Health Work* **2017**, *8*, 36–41. [CrossRef]

33. Driss, T.; Vandewalle, H. The measurement of maximal (anaerobic) power output on a cycle ergometer: A critical review. *BioMed. Res. Int.* **2013**, *2013*, 589361. [CrossRef]

34. Sado, N.; Yoshioka, S.; Fukashiro, S. Hip Abductors and Lumbar Lateral Flexors act as Energy Generators in Running Single-Leg Jumps. *Int. J. Sports Med.* **2018**, *39*, 1001–1008. [CrossRef]

35. Kariyama, Y.; Hobara, H.; Zushi, K. Differences in take-Off leg kinetics between horizontal and vertical single-Leg rebound jumps. *Sports Biomech.* **2017**, *16*, 187–200. [CrossRef] [PubMed]

36. Cho, K.H.; Beom, J.W.; Lee, T.S.; Lim, J.H.; Lee, T.H.; Yuk, J.H. Trunk muscles strength as a risk factor for nonspecific low back pain: A pilot study. *Ann. Rehabil. Med.* **2014**, *38*, 234–240. [CrossRef] [PubMed]

37. Tse, M.A.; McManus, A.M.; Masters, R.S. Trunk muscle endurance tests: Effect of trunk posture on test outcome. *J. Strength Cond. Res.* **2010**, *24*, 3464–3470. [CrossRef] [PubMed]

*Article*

# Agreement between Dribble and Change of Direction Deficits to Assess Directional Asymmetry in Young Elite Football Players

Athos Trecroci [1], Tindaro Bongiovanni [2], Luca Cavaggioni [1,3], Giulio Pasta [4], Damiano Formenti [5,*] and Giampietro Alberti [1]

[1]  Department of Biomedical Sciences for Health, Università degli Studi di Milano, 20122 Milano, Italy; athostrec@gmail.com (A.T.); cavaggioni.luca@gmail.com (L.C.); giampietro.alberti@unimi.it (G.A.)

[2]  Nutrition, Hydration & Body Composition Department, Parma Calcio 1913, 43044 Parma, Italy; tindaro.bongiovanni@gmail.com

[3]  Department of Endocrine and Metabolic Diseases, 20021 Milan, Italy

[4]  Medical Department, Parma Calcio 1913, 43044 Parma, Italy; ghitopasta@hotmail.com

[5]  Department of Biotechnology and Life Sciences (DBSV), University of Insubria, 21100 Varese, Italy

*  Correspondence: damiano.formenti@uninsubria.it; Tel.: +39-347-4128375

Received: 1 April 2020; Accepted: 5 May 2020; Published: 8 May 2020

**Abstract:** This study aimed to examine the agreement between asymmetries of dribble and change of direction (COD) deficits and to determine their potential difference to each other. Sixteen young elite football players were recruited and tested for sprint (over 10 m), dribbling ($90°COD_{dribbling}$) and COD ($90°COD_{running}$) performance in dominant (fastest) and non-dominant (slowest) directions. Dribble and COD deficits were computed to express dribbling and COD ability without the influence of acceleration. The asymmetric index (AI%) of both dribble and COD deficits were obtained for both directions. The level of agreement between dribble and COD deficits was assessed by Cohen's kappa statistic ($\kappa$). Results showed that AI% measured by dribble and COD deficits presented a poor level of agreement ($\kappa = -0.159$), indicating their imbalance did not favor the same direction. Moreover, AI% of the dribble deficit was significantly higher than those of the COD deficit. This study demonstrated that asymmetries in dribbling and change of direction performance (measured by dribble and COD deficit) were not in agreement to favor the same direction, also displaying a significant difference to each other. Practitioners should consider the task-specificity of asymmetry to reduce the imbalance in dribbling and COD performance.

**Keywords:** agreement; imbalance; football skills; football performance

## 1. Introduction

The combined asymmetrical and unpredictable nature of football prompts each player dribbling or changing direction in multiple directions (chaotically) within the pitch, which is unlikely to be equally distributed during a match [1]. Moreover, additional inherent factors (e.g., playing position, tactical constraints and players' leg or directional preference) may also contribute to influencing the players' movements within the pitch favoring predominantly their dominant side or direction to the detriment of the non-dominant one [2]. Although it would be advantageous for team sport athletes to express similar dribbling and change of direction (COD) performance toward different directions (right versus left) [3], they often manifest a certain degree of asymmetry, even throughout the season [1,4], that should be opportunely quantified.

Despite the apparent relevance of assessing dribbling and change of direction (COD) asymmetries, the available literature is scarce. Most of the studies used the completion time (the total time to cover a specific course) to detect the dribbling [5] and COD performance [2,6] for imbalance purposes. It has been previously observed that completion time might be biased by an individual's sprint capacity either via dribbling [7,8] or changing direction assessment [9,10]. To overcome this issue, dribble and COD deficits have been proposed to provide practitioners with a more valid and isolated measure within a field-based context, limiting the impact of acceleration [7–10].

A recent study quantified the directional asymmetry in the 505 COD test by deficit and total time between groups of different team-sport athletes (e.g., football, basketball and cricket). It was found that all athletes manifested a certain degree of asymmetry between the dominant and non-dominant side for both COD deficit and total time [11]. The authors also concluded that, being COD deficit unbiased toward individuals with higher acceleration capacity, its use should be preferred to compare asymmetry in respect of total time [11]. Moreover, in the studies of Dos'Santos et al. [3,12], the asymmetry of COD deficit reported higher percentages compared with total time with 35% of the subjects exhibiting values greater than the asymmetric threshold (14.5%) and 49% showing asymmetries greater than 10% [3], which has been previously used as a limit for an acceptable imbalance [12–14]. Although a dearth of research exists on COD deficit asymmetries, no information is available on the use of dribble deficit to quantify directional asymmetry.

Following the available literature, different assessments of asymmetry over a certain task may detect diverse levels of imbalances rarely favoring the same side or direction [15]. Madruga et al. [16] investigated whether the asymmetry was consistent between three unilateral jump-based tests in team sport athletes. The authors reported a low level of agreement suggesting that the asymmetries rarely favored the same dominant side. Similarly, Bishop et al. [17] reported slight to a fair agreement in the asymmetry within unilateral strength and jumping-based tests. Taken all together, these findings highlight the task-specificity of asymmetry, which should be considered when interpreting any performance influenced by leg or directional dominance in team sports. This may have important implications on the assessment of dribbling and COD asymmetries in football players. Dribbling and changing direction is pivotal to successfully compete in football. Quick and accurate change of directions while dribbling a ball allows a player to pass her or his opponent more easily, to invade a specific field area and to create a numerical superiority for increasing any chance of scoring a goal. In this context, besides quantifying the asymmetry using dribble and COD deficits, knowing whether there is consistency across them would be of practical importance. Of note, this may provide practitioners with useful information to target additional exercises for each individual's dominant (faster or preferred) and non-dominant (slower or non-preferred) side [17], which might differ between dribbling and CODs.

Therefore, the aim of the study was twofold: i) to examine the degree of agreement between dribble and COD deficit asymmetries in favoring the same direction; ii) to determine the extent of each dribble and COD deficit asymmetry and the possible difference to each other. Dribbling and COD are different movement tasks with the former more complex and technically demanding, especially concerning dominant and non-dominant sides [5]. Given the supposed task-specificity of asymmetry, we hypothesized that asymmetries of dribble and COD deficit would not favor the same direction, with the former displaying greater values than the latter.

## 2. Materials and Methods

### 2.1. Experimental Approach

In this cross-sectional study, 16 young football players from a professional club were tested for their dribbling and change of direction ability via a 90°COD test (for both dominant and non-dominant directions) over 10 m (with 5-m entry and exit). Dribble and COD deficits were employed to offer an actual ability to dribble or change direction without the influence of acceleration capacity. Then, the asymmetry index of dribble and COD deficit was computed to establish their level of agreement through the kappa coefficient. The derived asymmetries were also compared to each other to detect whether a potential difference would exist between dribbling and changing direction.

*2.2. Subjects*

Sixteen young elite football players from the same professional club (age 14.5 ± 0.8 years, body weight 64.3 ± 6.2 kg, height 177.1 ± 4.9 cm, maturity offset 1.05 ± 0.30 years) voluntarily participated in the study. The selected sample size was above the minimum value requested for conducting a Cohen's kappa agreement study [18]. All participants and their parents or guardians were informed about the purpose and potential experimental risks. After a deep description of the study, written consent was obtained from subjects and their parents or guardians to participate in the investigation. The study was approved by the Ethical Committee of the local Institution, in accordance with the Helsinki's declaration.

*2.3. Testing Procedures*

The subjects took part in the experimental procedure in June and were tested on an outdoor artificial turf at the same time of the day (i.e., from 3 p.m. to 5 p.m.). The subjects participated in two sessions. The first session involved a familiarization procedure in which all subjects gained confidence with the testing battery. Additionally, height, sitting height and body mass were taken by a stadiometer (SECA 213, Germany) and a portable scale (813, Germany) to the nearest of 1.0 cm and 0.1 kg, respectively. In the second session, a testing battery including 10-m sprint and 90°COD test (executed with and without a ball) was randomly arranged. A 5-min standardized warm-up based on forward and backward jogging, acceleration, deceleration and skipping movements up to 5 m, was employed before undertaking the first test [19]. An electronic timing gates system (Witty, Microgate, Bolzano, Italia) was used to record the total time for 10-m sprint, dribbling and COD performance with the gates set at 0.7 m above the ground. The foremost foot was placed 0.3 m behind the starting line.

2.3.1. Sprint Assessment

Each subject, when ready, sprinted over a 10 m from a two-point staggered stance. The subjects performed three maximal efforts interspersed by 2 minutes of passive recovery. The best performance time was considered in the analysis.

2.3.2. Dribbling and Change of Direction Assessment

A 90° change of direction test for dribbling (90°COD$_{\text{dribbling}}$) and running (90°COD$_{\text{running}}$) was employed. The layout of the test is shown in Figure 1. All players were instructed to perform three bouts with the ball and three bouts without the ball for each direction (right and left) with 2 minutes of passive recovery in between. The best performance of the three bouts (in each direction) was considered for subsequent analysis. The distance between the starting line to the cone and between the cone and the finish line was 5 m each. For 90°COD$_{\text{dribbling}}$, the players were requested to dribble the ball around the cone with a minimum of two touches (with the same foot) along each 5-m path. For 90°COD$_{\text{running}}$, they were instructed to change direction around the cone using the same side-step technique in each bout, to avoid any influence due to different COD execution technique. In case of hitting or touching the cone (even with the ball) at the turning point, the player was stopped and invited to repeat the bout after 2 minutes of recovery. The 90°COD$_{\text{dribbling}}$ and 90°COD$_{\text{running}}$ performance were initially measured by the total running time to complete the 5-m + 5-m course. Based on the recommendations of previous studies [3,9], we decided to employ a COD deficit for inferential analysis on asymmetry while using total running time for descriptive purposes. The dribble deficit was calculated by subtracting the 90°COD$_{\text{running}}$ total time from the 90°COD$_{\text{dribbling}}$ total time. The COD deficit was calculated by subtracting the 10-m sprint time from the 90°COD$_{\text{running}}$ total time. The fastest mean value between right and left directions was deemed as dominant (D) and the slowest mean value was considered as non-dominant (ND) [3].

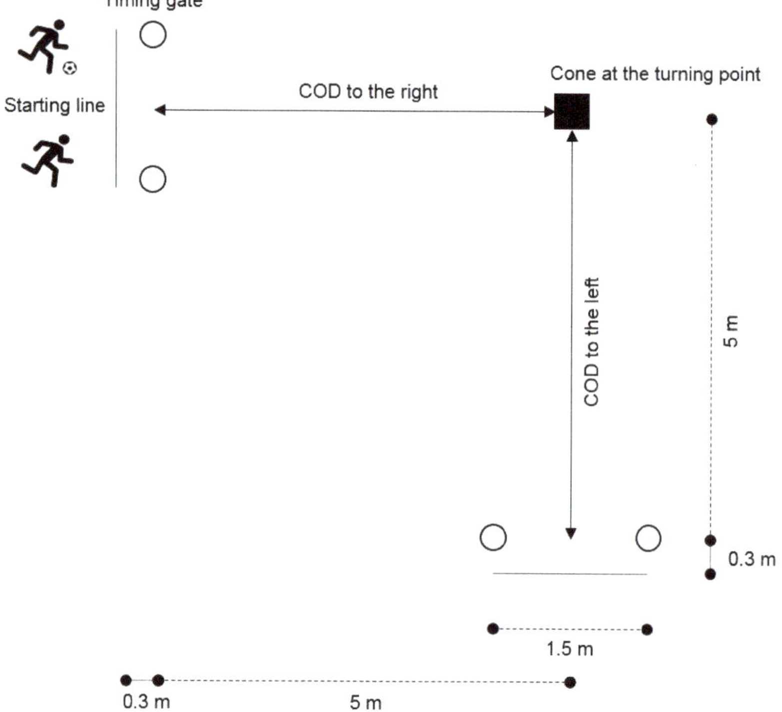


**Figure 1.** The layout of the 90° change of direction (COD) test. The black silhouettes with and without the ball identify the dribbling and COD performance, respectively.

### 2.3.3. Asymmetric Index Calculation

For both dribble and COD deficits the asymmetry index (AI%) was computed with the following formula:

$$\text{AI\%} = ((D - ND)/D) \times 100$$

Likewise, as previously proposed by Dos'Santos et al. [3], an asymmetry threshold (AT%) was also obtained to determine whether an individual can be considered as asymmetrical with the formula [3]:

$$\text{AT\%} = \text{AI\% mean} + (0.2 \times \text{SD})$$

where SD is the standard deviation of the AI% mean.

### 2.4. Statistical Analysis

The Shapiro–Wilk's test was conducted to verify if all data were normally distributed. The AI% resulted in non-normal distribution. Relative and absolute reliability was assessed for all tests using the intra-class correlation coefficient (ICC), standard error of the measurement (SEM) and the coefficient of variation (CV), respectively. Paired t-tests or Wilcoxon signed-rank test were used to detect differences between D and ND directions in 90°COD$_{\text{running}}$ (for total running time and COD deficit) and 90°COD$_{\text{dribbling}}$ (for total dribbling time and dribble deficit) tests, and between the AI% of dribble and COD deficit, respectively. The effect size of each difference was detected by Cohen's d ($d$) computation. The corresponding $d$ was classified as *trivial ($d < 0.2$), small ($0.2 < d < 0.6$), ($0.6 < d < 1.2$) moderate, ($1.2 < d < 2.0$) large, ($2.0 < d < 4.0$) very large* and *($d > 0.8$) near perfect.*

The degree of agreement between the AI% of dribble and COD deficit was assessed by Cohen's kappa statistic ($\kappa$). We used $\kappa$ coefficient as an appropriate tool for assessing the agreement of directional asymmetry (between the two tests) involving right and left dichotomous variables. The $\kappa$ coefficient described the chance-corrected proportional agreement determining how consistently an asymmetry in dribble and COD deficit agreed on the same direction [20,21]. Specifically, $\kappa$ was given by the formula:

$$\kappa = (\text{Observed Agreement} - \text{Chance agreement})/(\text{Maximum agreement} - \text{Chance agreement})$$

where the observed agreement defines the percentage proportion of the directions (right and left) for which dribble and COD deficit agree, and the chance agreement defines the overall random agreement probability that they agree on the same direction. According to Viera and Garrett [20], the following levels of agreement were considered: $\kappa < 0.00$ (*poor*), $0.01 < \kappa < 0.20$ (*slight*), $0.21 < \kappa < 0.400$ (*fair*), $0.41 < \kappa < 0.60$ (*moderate*), $0.61 < \kappa < 0.80$ (*substantial*) and $0.81 < \kappa < 0.99$ (*almost perfect*). Statistical analysis was performed using IBM Statistical Package for the Social Science version 21.0 (IBM Corp.; Armonk, NY, USA). An $\alpha$-value of 0.05 was set as a criterion level of significance. Ninety-five percent confidence intervals (95% CI) are shown in squared brackets. Data are reported as mean ± standard deviation (SD).

## 3. Results

ICC values showed excellent reliability in 10m sprint (ICC = 0.95, 95% CI [0.86 to 0.98]; SEM = 0.02 s, CV = 1.8%), $90°\text{COD}_{running}$ test for D (ICC = 0.93, 95% CI [0.88-0.96]; SEM = 0.03, CV = 2.3%) and ND directions (ICC = 0.94, 95% CI [0.86-0.95]; SEM = 0.03, CV = 2.5%), $90°\text{COD}_{dribbling}$ test for D (ICC = 0.88, 95% CI [0.61 to 0.96]; SEM = 0.097 s, CV = 3.3%) and ND directions (ICC = 0.88, 95% CI [0.66 to 0.96]; SEM = 0.105, CV = 3.5%). The descriptive statistics of each performance outcome with the inclusion of asymmetry are shown in Table 1. In $90°\text{COD}_{running}$ test, significant differences were observed between D and ND for total running time and COD deficit with *large* and *moderate* effects, respectively (p < 0.0001, d = −1.07, 95% CI [−1.84 to −0.30] and p < 0.0001, d = −0.73, 95% CI [−1.47 to 0.00], respectively). Likewise, in $90°\text{COD}_{dribbling}$ test, significant differences were observed between D and ND for total dribbling time and dribble deficit with *small* effects (p < 0.0001, d = −0.55, 95% CI [−1.28 to 0.17] and p < 0.0001, d = −0.57, 95% CI [-1.30 to 0.15], respectively). The Wilcoxon signed-rank test revealed a significant difference between AI% of dribble deficit and AI% of COD deficit (Z = −2.275, p = 0.021). The AT% of dribble and COD deficits were 17.22% and 41.62%, respectively.

**Table 1.** Descriptive statistics of performance outcomes.

| Physical Performance Tests | Mean ± SD | 95% CI |
|---|---|---|
| *10 m sprint* | | |
| Sprint running time (s) | 1.89 ± 0.09 | 1.83 to 1.94 |
| $90°COD_{running}$ | | |
| Total running time D *** [a] (s) | 2.43 ± 0.06 | 2.39 to 2.46 |
| Total running time ND (s) | 2.50 ± 0.07 | 2.46 to 2.54 |
| COD deficit D *** [b] (s) | 0.54 ± 0.09 | 0.49 to 0.59 |
| COD deficit ND (s) | 0.61 ± 0.10 | 0.56 to 0.67 |
| $90°COD_{dribbling}$ | | |
| Total dribbling time D *** [c] (s) | 2.84 ± 0.14 | 2.76 to 2.92 |
| Total dribbling time ND (s) | 2.93 ± 0.18 | 2.84 to 3.03 |
| Dribble deficit D *** [c] (s) | 0.33 ± 0.10 | 0.27 to 0.38 |
| Dribble deficit ND (s) | 0.43 ± 0.12 | 0.37 to 0.50 |
| *Asymmetry* | | |
| AI COD deficit (%) | −14.62 ± 13.03 | −21.56 to −7.67 |
| AI Dribble deficit * (%) | −36.11 ± 27.56 | −50.80 to −21.42 |

*** Significant (p < 0.0001) difference from ND, * Significant (p < 0.05) difference from AI COD deficit (%). [a] *Large* effect size d versus ND, [b] *moderate* effect size d versus ND, [c] *small* effect size d versus ND. Note: D = dominant, ND = non-dominant, COD = change of direction speed, AI = asymmetry index, SD = standard deviation, CI = confidence interval.

Figure 2 shows the individual data for dribble and COD deficit asymmetries. In 6 out of 16 players, the two AIs% favored the same direction with a resultant observed agreement of 0.38 (38%). The random probability agreement that dribble and COD deficit favored the right and left directions were ~39% and ~10%, respectively, with a chance-corrected proportional agreement of 0.46 (46%). The resultant κ score indicated a *poor* agreement of −0.159 (standard error = 0.187, 95% CI [−0.526 to 0.208]) between AIs% of dribble and COD deficits in favoring the same direction.

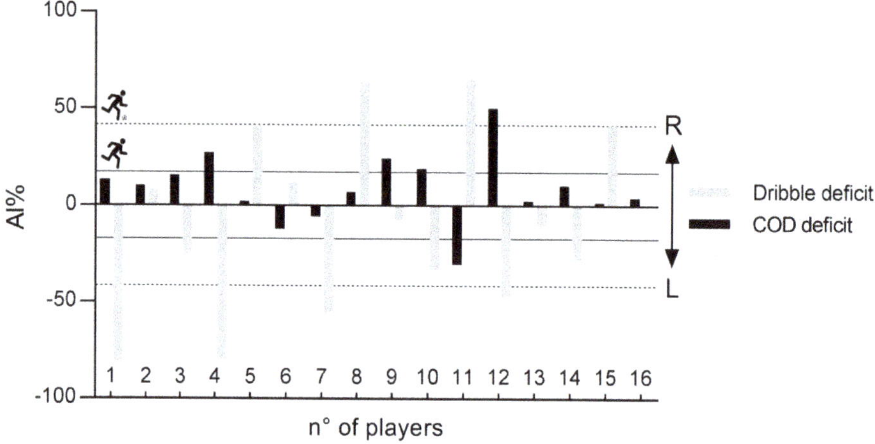


**Figure 2.** Individual asymmetry data (AI%) for dribble and change of direction (COD) deficits. Bars above the 0 line shows the asymmetry favoring the right direction, and bars below the 0 line shows the asymmetry favoring the left direction. The dotted lines indicate the asymmetry threshold of dribble deficit while the solid lines indicate the asymmetry threshold of COD deficit, respectively.

## 4. Discussion

The main finding of this study was that the asymmetry measured by dribble and COD deficits presented a *poor* level of agreement, indicating they did not favor the same direction. Moreover, it has been shown that, on average, AIs% of dribble and COD deficits were significantly different from each other, with the former presenting the highest values. These findings are in line with our hypothesis that dribble and COD deficits would not exhibit asymmetries favoring the same direction, with the former displaying the highest value.

Demonstrating whether (or not) the level of asymmetry (i.e., right versus left) is consistent across dribbling and COD performance provides practitioners with practical information that can be helpful to design targeted training strategies. According to the present results, while a player exhibited a fast change direction with the ball on a given side (right), she or he tended to display a fast change direction without the ball toward an opposite one (left). For example, Figure 2 shows only 6 out of 16 (~ 38%) players presenting an AI% favoring the same side, whereas most of the AIs% were not consistent across dribble and COD deficits. Indeed, the probability that asymmetries of both dribble and COD deficits would favor the same direction by chance was 46%, which is higher than the observed agreement. As such, the resultant level of κ score, which has the peculiarity of removing any agreement by chance, indicated that they did not produce similar results on a given side or direction. Taken all together, these findings also suggest that the asymmetry for $90°COD_{dribbling}$ and $90°COD_{running}$ tests over young subjects and using a common metric (e.g., deficit) is task-specific. This is supported by the study of Bishop et al. [17] in which the authors examined whether asymmetries were consistent across unilateral strength and common jumping-based tests (e.g., single-leg countermovement jump and single-leg broad jump) for peak force and impulse (eccentric and concentric). Most of the agreement for peak force (κ = 0.05) and impulse (−0.25 < κ < 0.32) ranged from *slight* to *fair* even across common tests, except for the *substantial* ($\kappa = 0.79$) agreement between single-leg countermovement jump and single-leg broad jump tests for concentric impulse. This provides evidence for the notion that asymmetry is task-specific. In fact, given the current results, practitioners should consider the task-specificity of asymmetry when interpreting dribbling and change of direction performance to implement targeted training strategies for an individual's dominant (faster or preferred) and non-dominant (slower or non-preferred) side [17]. For instance, it has been demonstrated that practicing with an emphasis on the non-preferred side (e.g., ND direction) by increasing accuracy and force in the kicks, ball control and speed may be a good practice to reduce asymmetry in dribbling [5]. Of note, the nature of these two motor actions (dribbling and COD) presents some peculiarities that differ from each other. Indeed, compared with COD, dribbling fast in multiple directions requires players a high technical (bilateral) proficiency to maintain the ball under control, which in turn slows their performance time.

To our knowledge, the present study is the first quantifying and comparing the dribbling and COD performance in young elite football players. Dribbling and COD are pivotal to successfully compete in youth football [22–24] with the former considered a field-based predictor of a player's success in one-to-one duels [25]. Unfortunately, some evidence exists on COD deficit in the current literature [3,9,26–28], there is a dearth of information on dribble deficit [7,8] and no data are available on its asymmetry. The unpredictability of game scenarios, together with inherent factors such as playing position, tactical constraints and players' leg or directional preference, may prompt players choosing predominantly their dominant side (at the expense of the non-dominant one) to address any football-specific maneuver [5]. The current results showed that the mean AI% of dribble deficit roughly doubled that of COD. The use of the ball requires players being able to perform complex movements depending on additional factors [29] (e.g., force, accuracy and precision kicking of dominant and non-dominant legs) that are likely to enhance the expected directional asymmetry among individuals [5]. Of note, in Figure 2, 6 ($\sim 38\%$) and 5 ($\sim 31\%$) out of 16 players were asymmetrical, presenting an AI% higher than the corresponding AT% for dribble and COD deficits, respectively. It is worth noticing that their values exceeded the common threshold of 10%, which has been previously used as a limit for an acceptable bilateral imbalance [12–14].

As regards COD deficit asymmetries, the present results appear in line with the study of Dos'Santos et al. [3], in which 35% of the subjects reported a significantly higher bilateral imbalance in COD deficit (by the 505 COD test) than the corresponding asymmetry threshold. However, according to recent studies, the relevance of any discussion about the thresholds and their capability to detect asymmetry has been questioned [4,15]. Bishop et al. [4,17] reported individual data for unilateral strength and jumping-based tests showing that asymmetries can sometimes be as large as 20%–40% with no bearing on a performance outcome (e.g., during CODs) [4]. Additionally, interpreting mean data without devoting attention to an individual approach would not depict a clear portrait of a player's asymmetry, and her or his training needs to reduce it. The current results appear to be in line with such consideration. Regarding dribble deficit, some individuals (e.g., player n° 1 in Figure 2) exhibited an imbalance higher than 50%, which is one and a half times larger than mean SD, while others were about on average (e.g., player n° 6 in Figure 2). Thus, it is evident how designing targeted training programs to the player n° 6 (with an AI% barely below the mean) would not likely contemplate the required additional exercises for the player n° 1 in an attempt to reduce the highest asymmetry. This information can be of practical relevance as practitioners are helped to plan any additional exercise on a more individual level to reduce asymmetry [17] in both dribbling and change of direction ability. It is notable that while COD asymmetries are detectable among team-sport athletes [11], within a homogeneous group as the present elite players, the inter-variability of dribbling and COD tests would limit the interpretation of the mean values. Bishop et al. [17] suggested to report and compare asymmetries to testing variability (e.g., CV%). In support of this, the inter-individual variability should be taken into account when attempting to screening young football players [30]. As such, it can be provided relevant information underpinning the monitoring and development of individualized or small-groups program routines [30].

This study presents limitations that should be acknowledged. The 90°COD test currently selected may be limited to represent the variety of dribbling techniques (close dribbling skills or a combination of long kicks and fast acceleration to run past an opponent) performed in matches [25]. Taking into account the task-specificity of asymmetry, further studies are warranted to examine how the evidence of no agreement with COD would be confirmed within a wider spectrum of dribbling skills. We also put in evidence that our findings cannot be surely extended also to other team sports. For instance, dribbling and COD abilities are determinants component in basketball. Thus, further studies are warranted to examine whether the current disagreement between dribble and COD deficit asymmetries in football would be found also in basketball players who dribble with upper limbs instead of lower limbs. Finally, we put in evidence that the present findings should be interpreted according to maturity status. Indeed, although the current players' maturity offset was fairly homogeneous, it might be possible that different results would come from heterogeneous maturity-related profiles, and consequently leading to different dribbling and COD deficit results.

## 5. Conclusions

This study demonstrated that asymmetries in dribbling and change of direction were not in agreement to favor the same direction, probably reflecting the different nature of these motor actions. As such, practitioners should consider the task-specificity of asymmetry to reduce the imbalance between dominant and non-dominant directions. For example, additional dribbling exercises placing the emphasis on the ND direction may represent a good strategy to improve the ND itself, without affecting the D direction. Of note, practitioners are encouraged to interpret asymmetry data with an individual approach to contemplate the required additional exercises for a given player to reduce her or his imbalance on a more individual level. In young elite players, assessing the direction of asymmetry during dribbling and changing direction appears pivotal to guarantee informative data on their potential individual imbalance. Finally, coaches and practitioners may benefit from data on players' directional imbalances to ameliorate both individual monitoring and training processes across the youth athletic development.

**Author Contributions:** Conceptualization, A.T. and D.F.; methodology, A.T., L.C. and T.B.; formal analysis, A.T. and D.F.; investigation, A.T., L.C., T.B., G.P. and G.A.; data curation, A.T.; writing—original draft preparation, A.T. and D.F.; writing—review and editing, A.T., T.B., L.C., G.P., D.F. and G.A.; supervision, G.A.; All authors have read and agreed to the published version of the manuscript.

**Funding:** This research received no external funding.

**Conflicts of Interest:** The authors declare no conflict of interest.

## References

1. Bishop, C.; Read, P.; Chavda, S.; Jarvis, P.; Brazier, J.; Bromley, T.; Turner, A. Magnitude or direction? seasonal variation of interlimb asymmetry in elite academy soccer players. *J. Strength Cond. Res.* **2020**, *1*. [CrossRef] [PubMed]
2. Rouissi, M.; Chtara, M.; Berriri, A.; Owen, A.; Chamari, K. Asymmetry of the modified Illinois change of direction test impacts young elite soccer players' performance. *Asian J. Sports Med.* **2016**, *7*, e33598. [CrossRef]
3. Dos'Santos, T.; Thomas, C.; Jones, P.A.; Comfort, P. Assessing asymmetries in change of direction speed performance: Application of change of direction deficit. *J. Strength Cond. Res.* **2019**, *33*, 2953–2961. [CrossRef]
4. Bishop, C.; Read, P.; Bromley, T.; Brazier, J.J.; Jarvis, P.; Chavda, S.; Turner, A. The association between interlimb asymmetry and athletic performance tasks: A season-long study in elite academy soccer players. *J. Strength Cond. Res.* **2020**. [CrossRef] [PubMed]
5. Teixeira, L.A.; Silva, M.V.; Carvalho, M. Reduction of lateral asymmetries in dribbling: The role of bilateral practice. *Laterality* **2003**, *8*, 53–65. [CrossRef] [PubMed]
6. Hart, N.H.; Spiteri, T.; Lockie, R.G.; Nimphius, S.; Newton, R.U. Detecting deficits in change of direction performance using the preplanned multidirectional australian football league agility test. *J. Strength Cond. Res.* **2014**, *28*, 3552–3556. [CrossRef] [PubMed]

7.  Ramirez-Campillo, R.; Gentil, P.; Moran, J.; Dalbo, V.J.; Scanlan, A.T. Dribble deficit enables measurement of dribbling speed independent of sprinting speed in collegiate, male, basketball players. *J. Strength Cond. Res.* **2019**, *1*. [CrossRef]

8.  Scanlan, A.; Wen, N.; Spiteri, T.; Milanović, Z.; Conte, D.; Guy, J.H.; Delextrat, A.; Dalbo, V.J. Dribble deficit: A novel method to measure dribbling speed independent of sprinting speed in basketball players. *J. Sports Sci.* **2018**, *36*, 2596–2602. [CrossRef]

9.  Nimphius, S.; Callaghan, S.J.; Spiteri, T.; Lockie, R.G. Change of direction deficit: A more isolated measure of change of direction performance than total 505 time. *J. Strength Cond. Res.* **2016**, *30*, 3024–3032. [CrossRef]

10. Nimphius, S.; Callaghan, S.J.; Bezodis, N.E.; Lockie, R.G. Change of direction and agility tests: Challenging our current measures of performance. *Strength Cond. J.* **2017**, *40*, 26–38. [CrossRef]

11. Dos'Santos, T.; Thomas, C.; Comfort, P.; Jones, P.A. Comparison of change of direction speed performance and asymmetries between team-sport athletes: Application of change of direction deficit. *Sports* **2018**, *6*, 174.

12. Dos'Santos, T.; Thomas, C.; Jones, P.A.; Comfort, P. Asymmetries in single and triple hop are not detrimental to change of direction speed. *J. Trainol.* **2017**, *6*, 35–41. [CrossRef]

13. Bishop, C.; Turner, A.; Maloney, S.; Lake, J.; Loturco, I.; Bromley, T.; Read, P. Drop jump asymmetry is associated with reduced sprint and change-of-direction speed performance in adult female soccer players. *Sports* **2019**, *7*, 29. [CrossRef] [PubMed]

14. Trecroci, A.; Formenti, D.; Ludwig, N.; Gargano, M.; Bosio, A.; Rampinini, E.; Alberti, G. Bilateral asymmetry of skin temperature is not related to bilateral asymmetry of crank torque during an incremental cycling exercise to exhaustion. *PeerJ* **2018**, *6*, e4438. [CrossRef]

15. Bishop, C.; Turner, A.; Jarvis, P.; Chavda, S.; Read, P. Considerations for selecting field-based strength and power fitness tests to measure asymmetries. *J. Strength Cond. Res.* **2017**, *31*, 2635–2644. [CrossRef]

16. Madruga-Parera, M.; Bishop, C.; Read, P.; Lake, J.; Brazier, J.; Romero-Rodriguez, D. Jumping-based asymmetries are negatively associated with jump, change of direction, and repeated sprint performance, but not linear speed, in adolescent handball athletes. *J. Hum. Kinet.* **2020**, *71*, 47–58. [CrossRef]

17. Bishop, C.; Read, P.; Lake, J.; Chavda, S.; Turner, A. Inter-Limb asymmetries: Understanding how to calculate differences from bilateral and unilateral tests. *Strength Cond. J.* **2018**, *40*, 1–6. [CrossRef]

18. Bujang, M.A.; Baharum, N. Guidelines of the minimum sample size requirements for Kappa agreement test. *Epidemiol. Biostat. Public Health* **2017**, *14*. [CrossRef]

19. Trecroci, A.; Cavaggioni, L.; Lastella, C.; Broggi, M.; Perri, E.; Iaia, F.M.; Alberti, G. Effects of traditional balance and slackline training on physical performance and perceived enjoyment in young soccer players. *Res. Sports Med.* **2018**, *26*, 450–461. [CrossRef]

20. Viera, A.J.; Garrett, J.M. Understanding interobserver agreement: The kappa statistic. *Fam. Med.* **2005**, *37*, 360–363.

21. Watson, P.F.; Petrie, A. Method agreement analysis: A review of correct methodology. *Theriogenology* **2010**, *73*, 1167–1179. [CrossRef] [PubMed]

22. Trecroci, A.; Longo, S.; Perri, E.; Iaia, F.M.; Alberti, G. Field-based physical performance of elite and sub-elite middle-adolescent soccer players. *Res. Sports Med.* **2019**, *27*, 60–71. [CrossRef] [PubMed]

23. Trecroci, A.; Milanović, Z.; Frontini, M.; Marcello, I.F. Physical performance comparison between under 15 elite and sub-elite soccer players. *J. Hum. Kinet.* **2018**, *61*, 209–216. [CrossRef] [PubMed]

24. Valente-dos-Santos, J.; Coelho-e-Silva, M.; Duarte, J.; Pereira, J.; Rebelo-Gonçalves, R.; Figueiredo, A.; Mazzuco, M.; Sherar, L.; Elferink-Gemser, M.; Malina, R. Allometric multilevel modelling of agility and dribbling speed by skeletal age and playing position in youth soccer players. *Int. J. Sports Med.* **2014**, *35*, 762–771. [CrossRef] [PubMed]

25. Wilson, R.S.; Smith, N.M.A.; de Ramos, S.P.; Giuliano Caetano, F.; Aparecido Rinaldo, M.; Santiago, P.R.P.; Cunha, S.A.; Moura, F.A. Dribbling speed along curved paths predicts attacking performance in match-realistic one vs. one soccer games. *J. Sports Sci.* **2019**, *37*, 1072–1079. [CrossRef]

26. Loturco, I.; Nimphius, S.; Kobal, R.; Bottino, A.; Zanetti, V.; Pereira, L.A.; Jeffreys, I. Change-of direction deficit in elite young soccer players: The limited relationship between conventional speed and power measures and change-of-direction performance. *Ger. J. Exerc. Sport Res.* **2018**, *48*, 228–234. [CrossRef]

27. Nimphius, S.; Geib, G.; Spiteri, T.; Carlisle, D. "Change of direction" deficit measurement in division I American football players. *J. Aust. Strength Cond.* **2013**, *21*, 115–117.

28. Taylor, J.M.; Cunningham, L.; Hood, P.; Thorne, B.; Irvin, G.; Weston, M. The reliability of a modified 505 test and change-of-direction deficit time in elite youth football players. *Sci. Med. Footb.* **2019**, *3*, 157–162. [CrossRef]

29. Duarte, J.P.; Tavares, Ó.; Valente-dos-Santos, J.; Severino, V.; Ahmed, A.; Rebelo-Gonçalves, R.; Pereira, J.R.; Vaz, V.; Póvoas, S.; Seabra, A.; et al. Repeated dribbling ability in young soccer players: Reproducibility and variation by the competitive level. *J. Hum. Kinet.* **2016**, *53*, 155–166. [CrossRef]

30. Nikolaidis, P.; Ziv, G.; Lidor, R.; Arnon, M. Inter-individual Variability in Soccer Players of Different Age Groups Playing Different Positions. *J. Hum. Kinet.* **2014**, *49*, 213–225. [CrossRef]

MDPI
St. Alban-Anlage 66
4052 Basel
Switzerland
Tel. +41 61 683 77 34
Fax +41 61 302 89 18
www.mdpi.com

*Symmetry* Editorial Office
E-mail: symmetry@mdpi.com
www.mdpi.com/journal/symmetry